Investigating Geography B

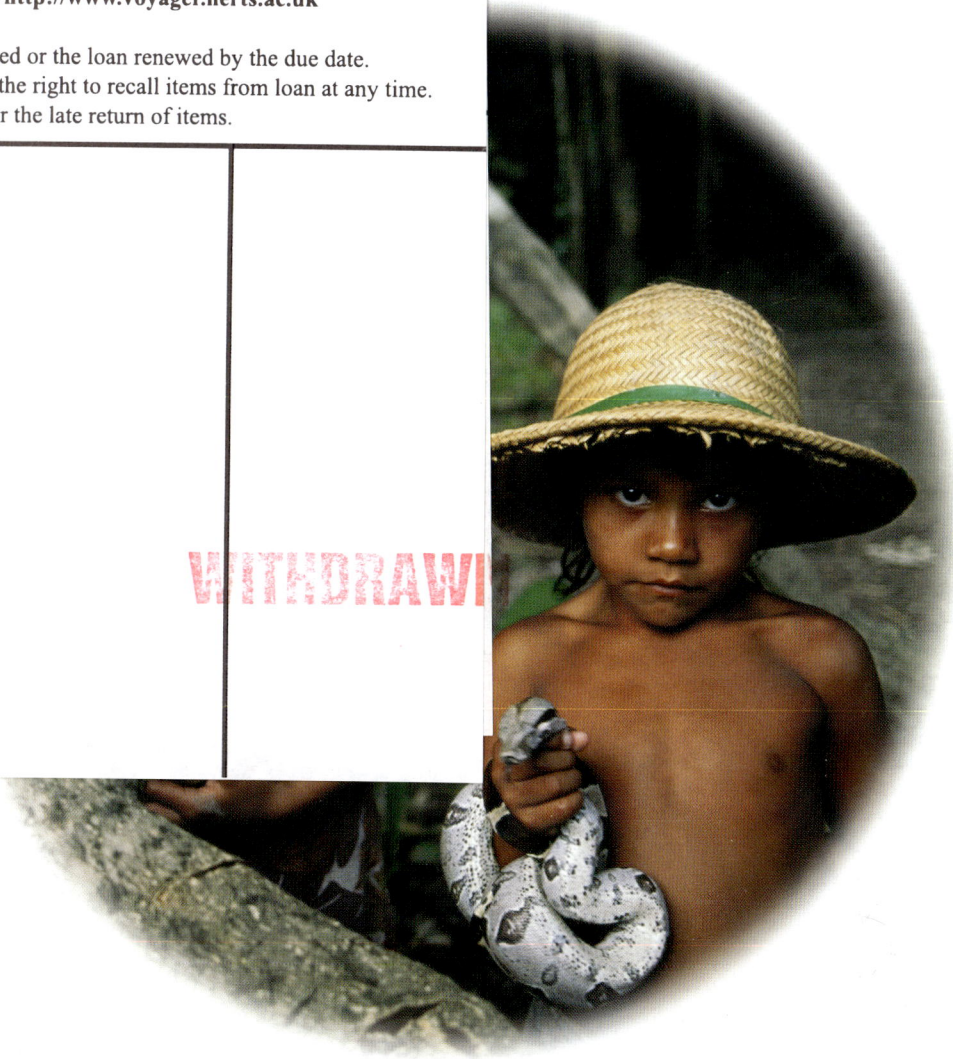

JACKIE ARUNDALE · SUE BERMINGHAM · SIMON CHANDLER · CHRIS DURBIN
GREG HART · BOB JONES · LINDA KING · FRED MARTIN · DIANE SWIFT
SERIES EDITORS: KEITH GRIMWADE & CHRIS DURBIN

Hodder & Stoughton

A MEMBER OF THE HODDER HEADLINE GROUP

Acknowledgements

The front cover illustration shows children in Brazil with pet snakes, reproduced courtesy of Stone Getty Images.
The publishers would like to thank the following individuals, institutions and companies for permission to reproduce copyright illustrations in this book: © American Stock/Hulton Archive, 3.5; © Andrew Ward/Life File, 1.15, 1.31r, 3.21; © Andy Crump/Still Pictures, 1.9; © Angela Maynard/Life File, 5.8; AP Photo/Dario Lopez-Mills, 1.24; AP Photo/Douglas Ende, 1.10; Birmingham Retail Crime Operation, 4.22; J Blakestone and Humberside County Council, 3.25, 3.37; © Bob Jones, 4.4, 4.6, 4.7, 4.11, 4.15, 4.16, 4.17, 4.18, 4.19, 4.30, 4.31, 4.32, 4.33, 4.34, 4.35, 4.37, 4.38, 4.43; © Bob Krist/Corbis, 3.17; © Dr B. Booth/G.S.F. Picture Library, 5.3, 5.4, 5.7, 5.28; © Bruce Coleman Collection/Colin Varndell, 3.11; © Bruce Coleman Collection/Granville Harris, 3.12; © Buddy Mays/Corbis, 1.16 (sloth); Natural Environment Research Council, 3.36; © Colin Taylor Productions, Knockholt, Kent, 6.9 (Germany); Corel, p1 (flag), p31bgc and r, 6.11b (farmer), 6.15 (3); © David A. Northcott/Corbis, 1.16 (harpy); Development Education Centre, Birmingham, 2.3, 2.4; The Home Office for extracts from Mirrlees-Black, C., Budd, T., Partridge, S. and Mayhew, P. (1998) The 1998 British Crime Survey, England and Wales.Home Office Statistical Bulletin 21/98. London: Home Office. (c) CrownCopyright.; Digital Stock, 3.23 (chef); Digital Vision, 4.13bg l, 6.15 (5), p123 (farmer); © Eye Ubiquitous/Corbis, 3.19; Eyewire, p5bg, p15bg, p20 (UN flag), p36 (flags), p80 (flags), 6.3 (bolts), 6.9 (flags);© Fiona Potter/Life File, 1.31l; © Galen Rowell/Corbis, 5.17br; Getty Images/Photodisc, 6.9 (UK); www.goldenjubilee.gov.uk, p4 (Queen); © Graham Burns/Life File, 3.3; © Greg Hart, 6.1b, 6.7a, 6.7b; Haydyn West/PA Photos, 4.2; by kind permission of the Home Office, p78 (Neighbourhood Watch logo); Hull Daily Mail, 3.35; Ikon Imaging, p4bg (Big Ben), p41bg (guards), 5.29r; Image Boss, p68bg, p74 (fence, flats); Infoterra, 2.19, 3.7, 3.28; Ingram Publishing, 6.3 (pizza, cheeseburger), 6.4 (pens, toothbrush), 6.9 (USA), p113 (fruit); James Davis Travel Photography, 1.23; © Jean Roche/Still Pictures, 5.9; © Jeff Griffin/Life File, 3.8, 5.14, 5.16; © Jeremy Hoare/Life File, 1.5, 1.8, 5.17bl, 5.17bc; © John Heseltine/Science Photo Library, 3.14; © John Maier/Still Pictures, 1.6, 1.33, 1.34; © Jonathan Blair/Corbis, 3.1; © Jorgen Schytte/Still Pictures, 2.2; © Kevin Schafer/Corbis, 5.22; © Klein/Hubert/Still Pictures, 5.10; © Landform Slides, 5.17tl; © Marc Garanger/Corbis, 3.10; © Mark Edwards/Still Pictures, 1.7, 5.31; Metropolitan Police, p74 (policeman);© Michael Ende, 1.29; © Michael Evans/Life File, 3.4; © Michael & Patricia Fogden/Corbis, 1.16 (ants); © Michael Sewell/Still Pictures, 1.16bg; National Portrait Gallery, p4 (Shakespeare); National Statistics, 4.41; © Nik Wheeler/Corbis, 4.28; © Nigel Shuttleworth/Life File, 1.4, 3.2; Copyright Nintendo Co., Ltd 2002, 5.35tr, 5.35br; Nomad, p4bg (TV), p5 (camera), 1.22 (car), p19 (paving, petition, laptop), p22bg, p27bg, p20 (flags), 2.12bg, p42bg, pp60–1bg, p72bg, p82 (computer), p89 (coin), 5.30, 5.35tl, p104t, p105bg, 6.3, 6.12 (lorries, clocks), p123 (rubbish, bin); PA Photos, p4 (Prince Charles); PA Photos/Abaca, 1.17; © Papilio/Corbis, 3.31; © Patrick Ward/Corbis, 1.2; Peak Pictures, 5.20; © Peter Dunkley/Life File, 5.11, 5.17tr; 3.25; Photodisc, p7bg, 1.12, 1.14, 1.16 (parrot, frog, monkey, spider ferns), p12bg, 1.22, p19 (sewer, people), p20bg, p21bg, 2.8, p28bg, 2.11bg, p31t and l, p36 (woman), p38bg, p41bg, 2.32, p43bg, p44bg, 3.6bg, 3.15 (pebbles), p57bg, 3.32, 3.39, 4.3bg, p67bg, p68 (camera), 4.12bg, 4.13bg c and r, p74 (keyboard), 4.26, pp82–3bg, p86bg, p88bg, p89 (goggles, glass), p90bg, p91bg, p95bg, p97bg, 5.29l, 5.35bl, p104 (banana), 6.3 (clothes, newspaper, bearings), 6.4, p109bg, 6.9 (people except USA, UK, Germany), 6.11a, 6.11b bg, 6.12, 6.15 (except 3 and 5), p117bg, p120bg, p123bg; Postojnska Jama, Turizem, Slovenia, 5.1, 5.24;© Robert Harding /Portland Sculpture & Quarry Trust - Art science heritage and education in the quarry environment. www.psqt@learningstone.net, 5.32; © Ron Gregory/Life File, 5.15; Scott Sinclair/Development Education Centre, Birmingham, 2.1; © Shai Ginotti/Corbis, 5.34; © Shehzad Nooran/Still Pictures, 6.2; © South West News Service, Bristol, 5.33; Steve Mitchell/Empics, 1.26; Steven M. Herrpich/Cincinnati Enquirer, 17/10/01, 4.1; © Sue Bermingham, 1.13, 5.26; © Ted Spiegel/Corbis, 1.3; Three Valleys Water, 6.3 (workman); Tony Marshall/Empics, 1.18; Tony Morrison/South American Pictures, 1.19; Trackair Aerial Surveys, 3.24; .www.tropix.co.uk (V. and M. Birley), 2.17; www.tropix.co.uk (Roland Birley), 2.18, 2.20; UpMyStreet.com, 4.42; Volta River Authority, Ghana, 2.30; © Water Aid/Jim Holmes, 6.14; © Water Aid and Tearfund, 2.24; World Travel Group, p8 (world cup). (b = bottom; t = top; l = left; r = right; c = centre; bg = background)
The publishers would also like to thank the following for permission to reproduce material in this book: 'Billions without clean water' by Claire Doole; information taken from BBC News Online; 'Private Water supplies "E coli risk"' information taken from BBC News Online; extract from The Cincinnati Enquirer, used with permission of The Cincinnati Enquirer; Nigel Dacre for his quotation used on ITN; © Development Education Centre for the Development Rose; this product includes mapping data licensed from Ordnance Survey® with the permission of the Controller of Her Majesty's Stationery Office. © Crown copyright. All rights reserved. Licence No. 100019872.
Every effort has been made to trace and acknowledge ownership of copyright. The publishers will be glad to make suitable arrangements with any copyright holders whom it has not been possible to contact.

Note about the Internet links in the book. The user should be aware that URLs or web addresses change regularly. Every effort has been made to ensure the accuracy of the URLs provided in this book on going to press. It is inevitable, however, that some will change. It is sometimes possible to find a relocated webpage by just typing in the address of the home page for a website in the URL window of your browser.

Orders: please contact Bookpoint Ltd, 130 Milton Park, Abingdon, Oxon OX14 4SB. Telephone: (44) 01235 827720. Fax: (44) 01235 400454. Lines are open from 9.00 - 6.00, Monday to Saturday, with a 24 hour message answering service.

British Library Cataloguing in Publication Data
A catalogue record for this title is available from the British Library

ISBN 0 340 80376 2

First Published 2002
Impression number   10 9 8 7 6 5 4 3 2 1
Year   2008 2007 2006 2005 2004 2003 2002

Designed and typeset by Nomad Graphique.
Printed in Italy for Hodder & Stoughton Educational, a division of Hodder Headline Plc, 338 Euston Road, London NW1 3BH.

Investigating Geography B

Contents

What do people think of the United Kingdom?

▽ **Figure 1.1** Lonely Planet
Guide to England

🏴 England

'Anyone who spends any extended period of time in England will sympathise with the locals' obsession with the weather, although in relative terms the climate is mild and the rainfall is not spectacular. The least hospitable months for visitors are November to February – it's cold and the days are short. March and October are marginal – there's more daylight but it can still be very cold. April to September are undoubtedly the best months, and this is, unsurprisingly, when most sites are open, and when most people visit. July and August are the busiest months, and best avoided if at all possible. The crowds on the coast, at the National Parks, in London and popular towns like Oxford, Bath and York have to be seen to be believed.'

The European Commission carried out a survey of young people across Europe to find out their views on countries in Europe. Each student was asked to write down the first five things that sprang to mind when they thought of a particular country. Below are the top five responses for the United Kingdom:

- **Shakespeare**
- **London**
- **BBC**
- **The Royals**
- **The Beatles**

Activities

1 List ten ways in which people can learn about other places, e.g. postcards, film locations, fiction, etc.

2 Draw a spider diagram of the UK **images** given on page 4.

3 Study the young people's responses to the European Commission survey. What information would you like to give to change their views of the UK?

4 Write down your view of the UK in 75 words.

How reliable are images?

When we are learning about a new place, we should not rely on one source of information because it may be biased. Every single photograph, whether taken by you or a professional photographer, has a purpose and a point of view. The person looking at your photograph may not see the same purpose and point of view.

▽ **Figure 1.2** Thatched cottage

Activities

5 Look at the three photographs of the UK. What do you think the photographer was trying to show?

6 What impression would a Brazilian student have if their only source of information was the photographs?

7 Do the photographs show how life really is in the UK in the 21st century?

8 Choose one photograph and write an interview with one of the people in the photograph. List the questions you would like to ask and their possible answers.

▽ **Figure 1.4** Morris dancers

△ **Figure 1.3** Welsh traditional dress

What images do we have of Brazil?

You may not like all the images portrayed of your country. Brazilians are also concerned about the images you have of their country.

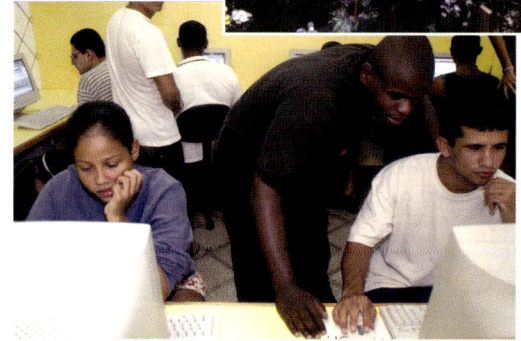

Figures 1.5 to 1.10

See page 23 for the answers to question 1.

Activities

1 Which of the photographs are images of Brazil? Which are images of other countries? Give reasons for your answer.

2 Which of the photographs match your image of Brazil?

3 Which of the photographs would you choose to support the following quotes:

> 'If travelling is your passion, Brazil is your destiny.' – Travel leaflet, Brazilian goverment

> 'All your dreams can come true in this land of happiness and hospitality.' – Travel leaflet, Brazilian goverment

> 'Sponsor a child today.'

> 'São Paulo, the great investment opportunity in South America.' – Governo do Estado de São Paulo

How do we change our images of a country?

A national survey in the UK in 1999 showed that over 80% of people learn about a **developing country**, like Brazil, from the television. Our news coverage of countries like Brazil falls into five categories:

- **Conflict**, war, **terrorism**
- Sport
- Western visitors to developing countries
- **Politics**
- **Natural disasters** or accidents

Documentaries and programmes like *Newsround* and *Blue Peter* can provide a good source of information and images about a country. But probably the best way to learn about a country is to visit it for ourselves. In 1997, ten students from the UK visited Brazil. They gave their impressions of the country before and after their visit (Figure 1.11).

Activities

4 The students who visited Brazil were asked to complete a word association exercise. Copy the list below and see how you get on with completing the same exercise.

Word Association

For each word in the list write down the first thing that comes into your head about Brazil:

- AMAZON
- INTERNET
- FOOD
- BRAZIL
- FOREST
- AIRPORTS
- GEOGRAPHY
- SHIP
- WILDLIFE
- NIGHT

5 Share your ideas with the rest of the class; are there any similarities?

Here are some of the answers the ten students gave before and after their visit.

	BEFORE	AFTER
FOREST	Rain, green, **canopy**, humid, fire, trees, intimidating	Dry, humidity, trees, darkness, monkey, wood, mosses, precious resource, green, amazing
BRAZIL	Amazon, Pelé, a shape rather than a word, location, heat, hot, adventure, **culture**	Hot, new age country, everything on our trip, variety, friends, very hot, memories, the trip, culture, diverse
WILDLIFE	Birds, animals, parrots, exotic, bush, spiders, insects, panda, worrying	None, animals, lack of, **scarce**, monkey, **reticent** but varied, insects, **sloth**, fascinating
NIGHT	Star, sleep, coldness, hot, sleeping, darkness, exciting	**Trek**, sleep, humidity, air conditioning, cold, time for **contemplation**, mosquitoes, stars, loud (in the rainforest)

△ Figure 1.11

Activities

Foundation

6 Use materials from local newspapers, tourist information offices, postcards, and travel brochures. Produce a leaflet to attract visitors to your local area.

Target

7 Produce a leaflet with the title 'The local area is an area of contrasts'. You may wish to take your own photographs or use existing materials such as newspapers, brochures, etc. Your leaflet should show both the advantages and disadvantages of living in your area.

Extension

8 You are to complete the Target activity to design a leaflet showing contrasts in your area. You should then write a letter to your local MP, explaining what you dislike about the area and offering your solutions to the problems.

Brazil – 500 years of history

FACT FILE: HISTORY OF BRAZIL

1500	**Pedro Alarves Cabral left Lisbon, Portugal in March and arrived in Brazil in April, declaring it a Portuguese colony**
1695	Gold discovered
1768	**Rio de Janeiro became the capital city, taking over from Salvador**
1822	Brazil given independence, ruler Pedro I
1889	**Brazil became a republic**
1958	Won World Cup football
1960	**Brasilia became the capital, taking over from Rio de Janeiro**
1962	Won World Cup football
1970	**Won World Cup football**
1985	End of military rule
1993	**Inflation at a record 2708%**
1994	Fernando Cardoso elected by landslide victory, bringing stability to the country. Won World Cup football.
2000	**500th birthday**
2002	Won World Cup football.

The first encounter between native South Americans and Europeans was recorded by Pero Vaz de Caminha, the official **scribe** of the Portuguese **flotilla** that accidently arrived on the coast of what is now Brazil on 22 April 1500. It started with an exchange of gifts.

[We] just gave them a red beret, a linen hood … and a black hat. And one of them gave us a head-dress of birdfeathers … another gave us a necklace of white beads.

Pero Vaz de Caminha

ICT link

www.brazil.com/menu/tools/html/about_brazil.htm has details on the five regions of Brazil.

http://news.bbc.co.uk/hi/english/world/americas/newsid_1231000/1231075.stm/ gives information on Brazil 1500 – 2000.

www.brazil.org.uk/index.php is the embassy page which is useful for general information on Brazil.

Activities

1 Imagine your town or village is twinning with a settlement in Brazil. What gifts would you suggest the Twinning Committee take to Brazil on their first visit?

What do you think Brazil is like?

Our images of a country can be very different to the reality.

It is difficult to imagine the size of the Amazon from Figure 1.12.

The River Amazon combines water from two sources. The River Negro is black coffee-coloured water because it contains acid from decomposed plants. The plants line the river from its warm source in Venezuela. The Solimões is milky coffee-coloured water; the water is colder because it begins in the Andes mountains. There is less plant life along the Solimões because of the temperatures. The water of the Solimões is more alkaline, as there are fewer decomposed plants. Figure 1.13 is taken from a boat on the Amazon – in the distance is an island in the river, not the other river bank.

The Brazilian Government is often criticised for cutting down and burning the rainforests. Environmentalists are concerned over the effects of global warming and the threat to wildlife as their habitats are being destroyed. In reality, many parts of the rainforest are being protected.

There are 36 National Parks in Brazil. These are fairly evenly distributed around the country, with six to eight parks in each region. There are also nature reserves and UNESCO World Heritage sites such as Lapinha and the National Park of Iguaçu.

The Amazon rainforest contains eight National Parks, seven biological reserves, ten ecological stations, three ecological reserves and 24 national forests. The Brazilian Institute for Space Research (INPE) constantly monitors the rainforest using satellites.

◁ **Figure 1.12**
The Amazon river flowing through rainforest

▷ **Figure 1.13**
The River Negro meeting the Solimões

△ **Figure 1.14**
Destruction of rainforest

▷ **Figure 1.15**
National Park in Brazil

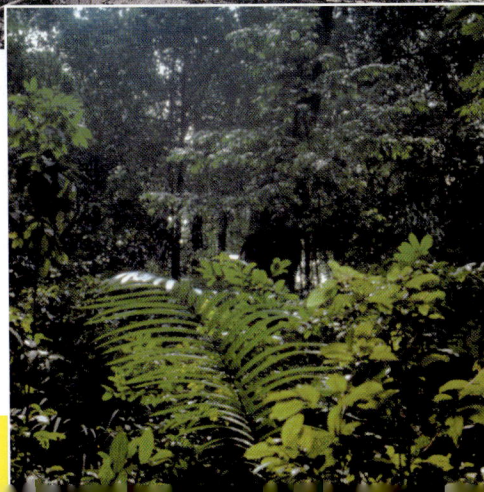

The Amazon Rainforest

Emergent layer
Insects, mosquitoes, swifts, whipporwill, night hawks, king vultures, eagles e.g. harpy, and flying geckos.

Interesting facts about the harpy
The top predator in the rainforest and the largest and most powerful of all eagles. The female has feet the size of a man's with sharp talons at the ends.

Canopy
Toucans, hawks, owls, pigeons, macaws, humming birds, parrots, sloth, lemurs, monkeys, flying squirrels and flying snakes.

Interesting facts about the sloth
Sloths have two or three toes. Their fur is tinted green from the algae that live on its coat. Many species of insects also live in its coat. The sloth spends most of its life hanging upside down from a tree, using its claws to grip.

Under canopy
Margay, ocelots, jaguars, deer, fruit bats, coati (racoon), tree frogs, geckos, and chameleons.

Interesting facts about the tree frog
The tree frog is well designed for life in this layer of the forest. Each finger or toe has a sticky suction cup that helps it to grip the trees as it climbs.

The floor
Beetles, flies, butterflies, mosquitoes, wasps, termites, ants, anaconda (world's largest snake) and the largest rodent (capybara), armadillos, anteaters (with a sticky 60 cm tongue) and frogs.

Interesting facts about the termites
A termites' nest (a termitary) can be over 2 m high and can contain 10 million termites. One million termites can eat 12 tonnes of dead wood in a year.

Activities

1 Trekking through the rainforest.
Imagine a walk through the rainforest. What sounds and sights would you experience?

△ Figure 1.16

Where is Brazil?

BRAZIL

Total area:
8 511 965 sq km

Population: 172 million

Currency: Real

Capital city:
Brasilia (pop 1.59 million)

Geographic coordinates:
10° 00' S, 55° 00' W

Comparing area:
Size of Europe

Total land boundaries:
14 691 km

Coastline: 7 491 km

Climate: mostly tropical,
but **temperate** in south

Famous Brazilians

In the UK we are aware of the many famous Brazilians, particularly in the world of sport. Have you heard of these people shown here?

△ **Figure 1.18** Rivaldo

▷ **Figure 1.17**
Gisele Bundchen

Activities

2 **a** Brazil is the largest country in South America and shares common boundaries with every South American country except Chile and Ecuador. Find out the names of the ten countries that border Brazil.

b Using your answers to question 1, produce a location map for Brazil. Using an atlas to help you, name and locate:
• two oceans
• one sea
• 13 countries in South America
• the capital city of Brazil
• the Equator
• the Tropic of Capricorn.

3 Use resources like the Internet and atlases to find comparison data for a UK Fact File.

4 The United Kingdom only has a land border with one other country. Can you name it?

Who lives in Brazil?

The original inhabitants of Brazil were the Indians. When the Portuguese arrived, there were an estimated 5 million Indians in Brazil, organised into 1 000 **tribes**.

Between 1540 and 1850, up to 8 million West Africans were brought to Brazil, as slaves. The slaves worked in the sugar cane industry. **Slavery** was finally abolished in 1888.

During the 19th century, the largest group of immigrants came from Portugal. An immigrant is a person who arrives to live in a new country. The Portuguese were followed by Italians and Lebanese but not in such large numbers. These people came to work on the **plantations**, in the rubber industry and in gold mining.

In the first half of the 20th century, **migrants** from Europe arrived in Brazil to escape war and economic problems. In 1908, 640 Japanese migrants came to Brazil. This started a trend and by 1969, a total of 247 312 had arrived. This group is the largest community of Japanese outside Japan.

Activities

1 You will need a base map of the world. Your map is going to show where people in Brazil originally migrated from.

For each group of people you will draw an arrow from their country of origin to Brazil. Label the countries on your map, showing where the migrants came from.

Arrows from Africa (Benin, Nigeria, Angola)

Arrows from Europe (Portugal, Italy, Germany, Spain, Poland, Lebanon)

Arrow from Asia (Japan)

▽ **Figure 1.19** Japanese quarter, São Paulo, Brazil

THE PEOPLE:	THEIR LANGUAGE:
55% European descent	Portuguese – Brazil is the only Portuguese-speaking country in South America
39% Mulatto	
5% African descent	
0.5% Japanese	**THEIR RELIGION:**
0.5% Others (according to the 1991 census)	90% Roman Catholic

People today

The majority of people in Brazil today are of European descent. 39% of people are **dual heritage** (mulatto), as their parents are from different ethnic groups. Brazil is a young country; 62% of its people are under 29 years old and 44% under 19 years old.

There are now only about 350 000 Indians in Brazil, members of just 210 tribes, according to the government's Indian agency. There are an estimated 50 tribes that have never had contact with the outside world. Most of the Indians live simple lives within the forest, with few facilities such as education and healthcare. The average life expectancy of traditional Indians is 42.6 years.

No other country outside the continent of Africa has such a large black population. However, the black people rarely hold positions of power in the government, the legal system and the armed forces. Even in Salvador, which was the capital and major slave port for nearly 300 years, where blacks make up more than 80% of the population, very few blacks are to be found in government.

Comparing population density in Brazil with the UK (national scale)

The world's population is unevenly spread. Some countries are very crowded (densely populated) and others have few people (sparsely populated). Which country, Brazil or the UK, do you think is the most crowded?

Population distribution within Brazil (regional scale)

Brazil has five main **regions** (Figure 1.21). The population is not spread out evenly. Some areas are more crowded than others.

Activities

2 Calculate the **population density** for both Brazil and the United Kingdom by dividing the total population by the total area.

	BRAZIL	UNITED KINGDOM
Area:	8 511 965 km²	244 111 km²
Population:	172 000 000	59 000 000

△ **Figure 1.20**

Your answer should be given in 'people per square km'.

3 Illustrate your answer to question 1. You could place dots inside a square which represents a square km (e.g. 5 cm x 5 cm). Each dot drawn inside the square represents one person per sq km. You might be able to think of a better method.

4 Using your answers to questions 2 and 3, which country is the most crowded?

5 **a** Study the map of population distribution in Brazil's five regions.
b Describe the pattern of where Brazil's population lives.
Hint: Coastal / Inland
             North / South
             East / West

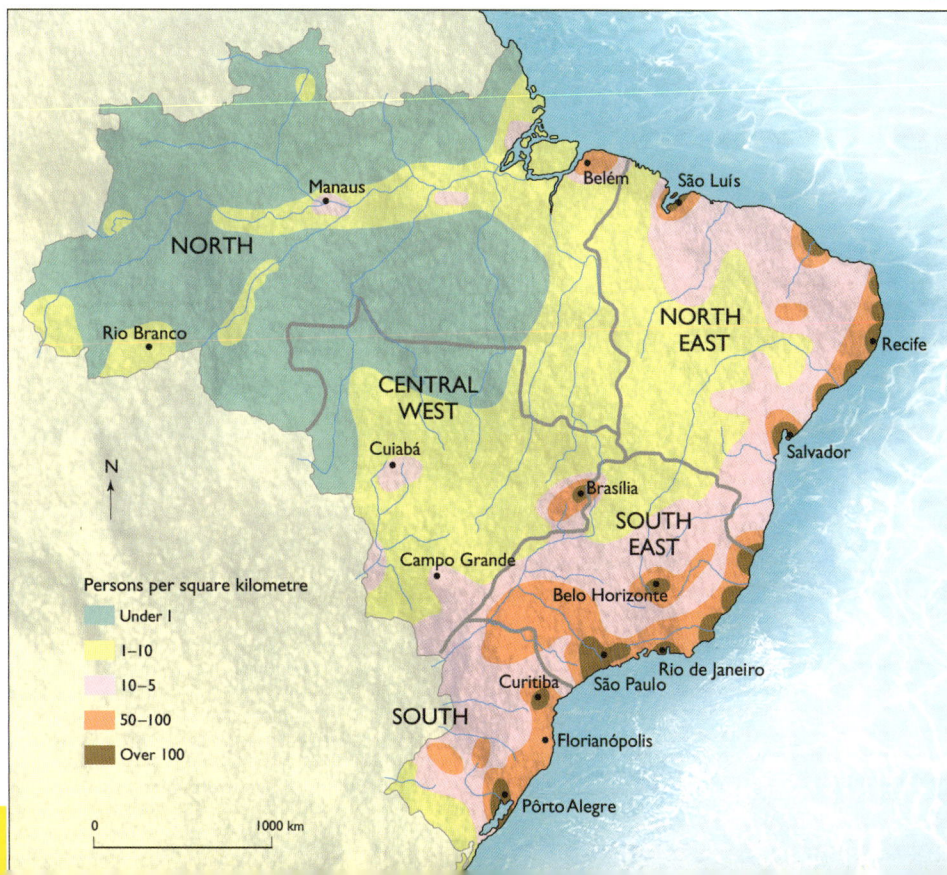

▷ **Figure 1.21** Map of Brazil's population distribution

Persons per square kilometre
- Under 1
- 1–10
- 10–5
- 50–100
- Over 100

Manaus, Belém, São Luís, NORTH, Rio Branco, Recife, NORTH EAST, CENTRAL WEST, Cuiabá, Salvador, Brasília, SOUTH EAST, Campo Grande, Belo Horizonte, Curitiba, São Paulo, Rio de Janeiro, SOUTH, Florianópolis, Pôrto Alegre

0   1000 km

What is Brazil's place in the world economy?

Brazil has the most successful economy in South America and is increasing trade with the rest of the world. 42% of Brazil's workforce is employed in the service industry, 31% is employed in agriculture and 27% in manufacturing industry. Manufacturing industry can also be called **secondary industry**. Secondary industries make new products from processing raw materials. Secondary industry makes up 36% of Brazil's **Gross Domestic Product (GDP).**

Brazil can use its own raw materials such as **bauxite** (containing aluminium), gold, iron ore, manganese, nickel, phosphates, platinum, tin, uranium, petroleum, **hydro-electric power (HEP)** and timber to make new products.

▽ **Figure 1.22** Secondary industries process **raw materials**

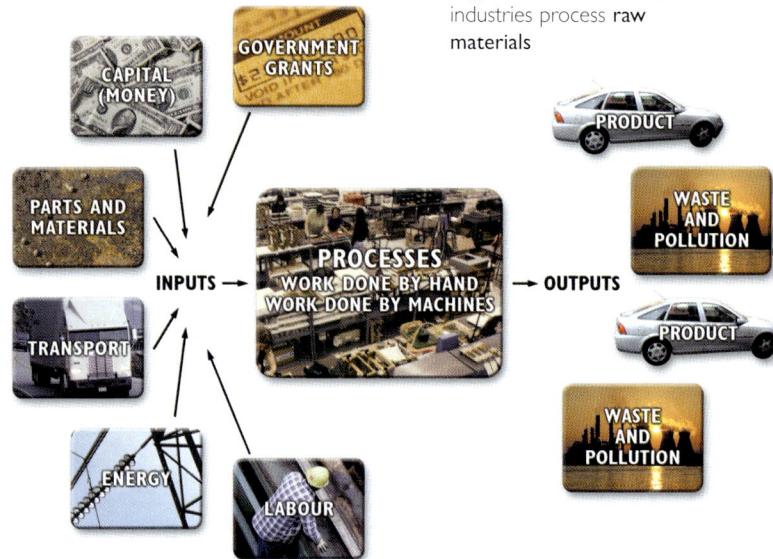

◁ **Figure 1.23** HEP plant, Brazil

▷ **Figure 1.24** Car manufacturing plant

The main industries in Brazil are textiles, shoes, chemicals, cement, tin, steel, aircraft, motor vehicles, soya beans, orange juice, beef, coffee, sugar. Brazil has transnational corporations such as Odebrecht, which has invested in 14 overseas countries and earned US$ 438.5 million profit in 1999.

In Brazil, an increasing percentage of people are employed in manufacturing and manufacturing production is growing. Also, manufacturing is making a bigger contribution to GDP and **exports** are increasing. This rapid growth in manufacturing means Brazil is classed as an **NIC**, a **Newly Industrialising Country**.

Gross Domestic Product

Figure 1.25 includes two figures that can be used to measure the wealth of a country – GDP and GDP per head. GDP is the total value of a country's domestic **economic output** in one year. GDP does not include any income from abroad. Another figure that is often used by geographers to compare wealth is GNP (Gross National Product) which does include income from **overseas investments**.

	BRAZIL	UNITED KINGDOM
Gross Domestic Product (GDP)	US$595.5 billion	US$1.36 trillion
GDP (per head)	US$9 500	US$22 800

△ **Figure 1.25** GDP for Brazil and UK (2000)

Activities

1 Find out the GNP for Brazil and the UK.

2 Why might some geographers choose to use GDP per head rather than the total GDP when comparing countries?

Brazil trades with all parts of the world, mainly exporting to the EU and the USA and mainly importing from the Middle East and USA. It has had a **trade surplus** since the early 1980s, earning more from its exports than it does from its imports.

Brazil is a member of two main trade organisations – LAFTA (the Latin America Free Trade Association) and MERCOSUL (Paraguay, Uruguay, Argentina).

Brazil's links

Brazil is the UK's largest trading partner in Latin America. Most of the UK's internationally active companies have links in Brazil.

Brazil is not only a manufacturing country. The **tertiary industry** in Brazil is also growing and making money, accounting for 50% of the GDP. The tertiary industry provides a service, including employment in the police, education, armed services, banking, health, government. Football, Formula One and tourism are three of the more famous tertiary industries in Brazil.

British Gas Glaxo
HSBC Reckitt and Colman
Lloyd's TSB
SmithKline Beecham
Pilkington ICI

As Brazil develops, it needs to generate an increasing amount of electricity for homes and industry. The government is overseeing a US$26 billion project to install an extra 25 000 **megawatts** of electricity generating capacity between the years 2000 and 2005.

In 1998 a huge oilfield was discovered off the coast of Rio de Janeiro. An oilspill in January 2000 in Guanabara Bay (Rio de Janeiro) has led to the company responsible spending over US$1 billion in additional environmental preventative measures to regain public support. Environmental centres have been built containing all the tools necessary to deal with oil spills around the Rio de Janeiro bays. Hundreds of **buoys** are located in the water which can detect oil and alert the centres.

△ **Figure 1.26** Formula One racing

Activities

3 Do less developed countries export more manufactured goods or more primary goods? Give reasons for your answer.

4 What does Britain export? Copy and complete the table below. You may need to carry out some research.

BRITAIN'S EXPORTS

PRIMARY GOODS	MANUFACTURED GOODS

Is there inequality in Brazil?

▽ **Figure 1.27** Lorenz Curve

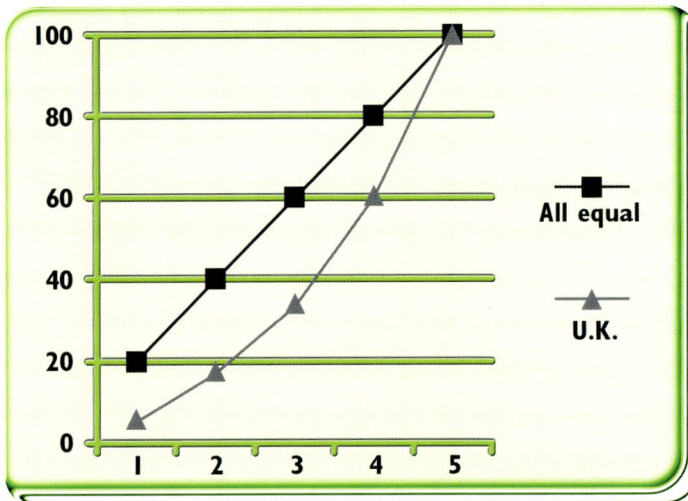

Within any country, there are some people who are very rich and others who are poor. Geographers can use a Lorenz curve graph to show how the wealth of a country is shared out amongst the population. In an ideal world, the wealth would be shared equally and the graph would be a straight, diagonal line. If you look at the graph (Figure 1.27) the horizontal axis shows the poorest person on the left to the richest person on the right. The vertical axis shows no money at the bottom and 100% of the wealth at the top.

› How to draw your own Lorenz curve

1. Draw a square 10 cm x 10 cm.
2. Mark off each cm along the horizontal and vertical axis.
3. Number each axis from 0 to 100%, 0 is in the bottom left corner.
4. Label the horizontal axis Population and the vertical axis Percentage of Wealth.
5. Plot the 'All equal' line, where the horizontal axis equals the vertical axis, e.g. 20% of the people have 20% of the wealth, 40% of the people have 40% of the wealth, etc.
6. Plot on figures for Brazil and the UK, using different colours for each line (see Figure 1.28).

› How to understand your Lorenz curve

If everybody in the country had the same amount of money, the graph would be a straight, diagonal line. Look at the space between the line for the UK and the 'All equal' line – the space shows how unequally the wealth is shared. In the UK the richest 20% have almost 40% of the wealth.

% of people		BRAZIL (% of wealth)	UK (% of wealth)
poorest	20	2	7
	40	7	18.5
	60	16.4	35.5
	80	33.4	60.3
	100	100	100

△ **Figure 1.28** How fairly is wealth shared within a country?

Activities

1. Look at the line you have drawn for Brazil. Choose the correct statement from below:
   - Brazil's wealth is shared more equally than the UK's.
   - The UK's wealth is shared more equally than Brazil's.

Brazil is considered to be one of the most unequal countries in the world and the difference between the richest and poorest people is increasing. Forty years ago, the richest 10% of people in Brazil had 34 times more income than the poorest 10%. Ten years ago, the richest people were earning 78 times more money.

△ **Figure 1.29** A helicopter in São Paulo in Brazil, which has more helicopter flights than any other city

Alex Bellos writes, 'Every morning the man leaves his gated community by helicopter. He is not alone: the sky is full of helicopters taking the rich to work above an endless sprawl of high-rise blocks and streets jammed with traffic…

According to the São Paulo Pilots' Association, more helicopter flights are made here than in any other city…

Well-off Paulistanos use their helicopters not only for commuting but also for travelling to their weekend beach homes, and even for shopping.'

△ **Figure 1.30** Article from the *Guardian*, 7 August 2000

About 50 million Brazilians are living in misery. The Brazilian Government use the word 'misery' to describe somebody who does not have enough income for food, clothing and housing. Many have to survive on a minimum wage of $77 per month. Within Brazil, the **rural** areas have a higher percentage of poor than **urban** areas.

▽ **Figure 1.31** Contrasting housing in Brazil

Is there inequality in Brazil?

Thirty-two million people in Brazil do not have access to clean water and 24 million are unable to read and write. Many people live in large shanty towns called 'favelas'. A favela is a town of shacks built by families out of materials found lying around. The poor have few opportunities for education, jobs and health care.

A geographer carried out research in Brazil in 1998 and used these adjectives to describe people living in favelas:

- creative
- knowledgeable
- resourceful
- active
- resilient

1 Write five sentences, each containing one of the five adjectives above. Your sentences should describe the people living in favelas and their lifestyle.

Case Study of Maria Cursi (a favela)

This favela has existed for over 40 years and is home to over 400 families. The favela is very crowded. Many of the houses have been there for a long time and have been made more permanent by the families. Today, over 80% of the houses in the favela are made from stone. There is no sewerage but it does have water and light.

The favela is on the edge of São Paulo. The people who live in the favela may have jobs in nearby factories, shops and offices. Some people are street sellers and others run small businesses from their home. Some of the original residents have now retired.

Many of the people living in Maria Cursi like it there. In 1995, the residents read in the newspaper that their favela was going to be demolished and that they would be re-housed in government flats. They started a campaign, with the help of MDF (Movement for the Defence of the Favela dweller). The residents carried out research on other favelas and government housing schemes; they met with officials and gathered information before voting on their future. Having considered all the evidence, 425 of the 432 families living in the favela voted and 409 were against demolishing their favela.

'Of course we want a better life. All of us here want that, but how? There's no shortage of co-operation. Neighbours and friends all help each other. But even with the help of others, you can only build a brick house if you have the money. For me it's impossible. What I earn doesn't even fill our stomachs.'

△ **Figure 1.32** Statement from Anna Lucia Florisbela dos Santos, favela resident

VIEWS OF THE RESIDENTS

All homes would be destroyed but there was no guarantee people would be re-housed.

All businesses would be destroyed.

To gain a government flat you would have to have registered employment and a regular salary.

They would no longer be designer of their own home.

They were worried about future rent increases and evictions.

The retired residents were especially worried.

Activities

The residents of Maria Cursi had ideas on how their lifestyle could be improved:

Mains sewerage

Paved streets

Involvement in decisions about the future of the favela

Foundation

2 Imagine you live in the Maria Cursi favela.

a Write a paragraph to show why you enjoy living there.

b Design a poster to stop your favela being demolished.

c Write a letter to the Brazilian Government, listing five ways in which your favela could be improved.

Target

3 Imagine you are a journalist for a São Paulo newspaper. You have been asked to write an article about the conflict over the future of Maria Cursi.

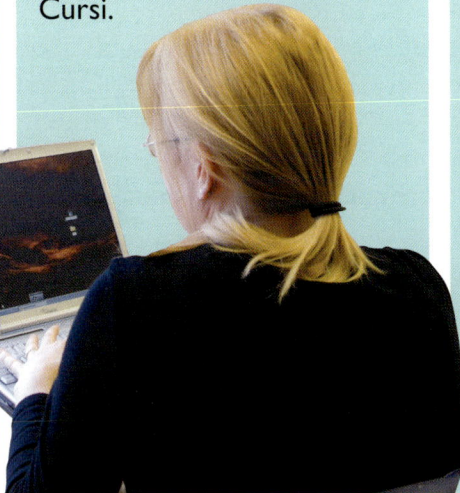

Your article should include:

a The history of the favela and a description of what it is like today.

b Artists' impressions of the proposed change to the favela (five- to eleven-storey high flats).

c An interview with Anna Lucia Florisbela dos Santos.

Extension

4 A planning enquiry is often held when people disagree over how a piece of land should be used. At the enquiry, both sides can put forward their views and a Government Inspector makes a final decision based on all the evidence.

a Write a persuasive speech for the Project Manager, explaining why the favela should be demolished and what are the advantages of building high-rise (five- to eleven-storeys) flats that have a living room, kitchen, utility room and bathroom.

b Write a response from the MDF, explaining why the people of Maria Cursi wish to stay in their favela.

STOP! DON'T DEMOLISH OUR FAVELA

What about the future?

Every year we are becoming more aware of how fragile our planet is and how human activities can damage it. Many countries are now working towards **sustainable development**. This means meeting our present-day needs without compromising the ability of future generations to meet their needs.

In 1989 the United Nations held a conference on Environment and Development. Brazil offered to host the next conference in 1992 in Rio de Janeiro. After that conference, governments worked on a document that 179 states could agree to. The final document was called **Agenda 21**, a $600 billion programme to save our planet in the 21st century.

Deforestation is still a current global environmental issue. Cutting down and burning trees destroys the habitats of a huge number of plants and animals, as well as threatening some of them with extinction. One attempt to reduce these problems in Brazil was the signing of an environmental crime bill by President Cardoso in September 1999. The bill states that deforestation is a crime punishable by stiff fines and jail sentences.

Crime bills were also signed for pollution. Air and water pollution in Rio de Janeiro, São Paulo, and several other large cities are of major concern to the Brazilian Government.

Various parts of Brazil have also been affected by mining, including the Amazon rainforest, causing water pollution and **land degradation** where the activities are not carefully managed.

Brazil takes an active role in reducing global environmental worries and has been involved in many **initiatives** and agreements on a range of issues including Antarctica, climate change, **desertification**, endangered species, nuclear testing ban, the ozone layer protection and whaling.

Not only must countries take care of the natural environment but we also need to consider the future of our cities. A sustainable city grows in a way that meets the needs of the present without reducing the quality of life for future generations.

A sustainable city might have:
- A transport system which reduces pollution and the number of cars on the road
- A range of housing types and prices to suit all needs and all budgets
- Land use planned to avoid long distance commuting
- Good **infrastructure** (roads, railways etc.) and services including schools and hospitals
- More open green spaces, such as parks.

Is Curitiba a sustainable city?

Curitiba is in the Paraná state of southern Brazil (25.5° South and 49° West) and is approximately 497 miles (800 km) south of Rio. Curitiba is one of the oldest cities in Brazil, founded in 1653 by the Portuguese. It is a modern city, with a population of almost 2 million. During the last elections the residents even voted using touch-sensitive screens.

Curitiba is 'world-famous for its imaginative and efficient transport system, its approach to waste management, its extensive parks, and its commitment to integrating environmental issues in all levels of education.' (Peter Hanneberg, *Enviro Magazine,* December 1992.)

There are over 500 000 cars in Curitiba, the highest car density in Brazil, and yet there are few traffic problems. The main reason for this is that few people use their cars because the bus service is so efficient. Reducing the use of cars has cut fuel consumption and improved air quality.

› Curitiba's public transport

Buses in Curitiba carry around 1.5 million passengers a day. Buses are colour-coded according to their route, so they are easy to identify. Red are for express services; green for inter-district routes; and yellow for feeder buses.

Passengers queue in long tubular bus shelters, enabling them to quickly board and alight the express services. The infrastructure is designed to give priority to public transport and bus drivers have control of traffic lights at junctions. Each of the five main roads has two middle lanes reserved for buses.

› Waste management

Curitiba has successfully organised the management of urban problems like waste.

▷ **Figure 1.34** Skyscrapers in Curitiba

Curitiba's Institute of Urban Research and Planning (IPPCU) has been responsible for the city's high environmental quality. The 'Rubbish That Is No Rubbish' scheme organises the recycling of waste. Residents sort waste into paper, cardboard, metal, plastic and glass. Trucks collect the sorted waste, once a week, and sell it to recycling companies.

› Curitiba's parks

Curitiba is one of the 'greenest' cities in the world, with around 22 million square metres of green open space. Many of the parks have small lakes and cycle-ways. Bus journeys to the parks are free at weekends. The largest park is the 'Parque Barigui', named after the river that runs through it.

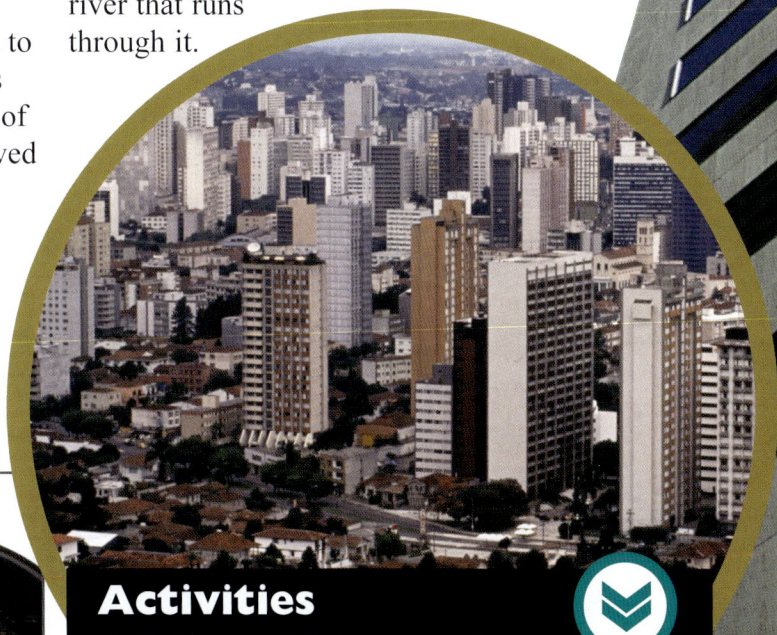

Activities

1 Use the information and ideas about Curitiba and knowledge about towns and cities near your school. Produce a checklist for planners of new towns and cities like Palmas (see below), so that they are healthy, safe places in which to live.

Palmas, 600 km north of Brasilia, did not exist 12 years ago. Today, Palmas is Brazil's fastest-growing city, with a population of 150 000 people.

△ **Figure 1.33** Tubular bus shelter, Curitiba

Assessment tasks

TV Directors have to make sure their programmes match certain **policy** criteria if they are to be shown in the UK.

THE PROGRAMME MUST:

Attract an audience

Have good reviews from critics, e.g. in newspapers

Be affordable

Be interesting

'Our news review showed that people want foreign news as long as we tell them why a far-away story matters and how it links to them. They want solutions rather than problems.'
Vin Ray, BBC

'People don't distinguish between domestic and foreign viewers – but between what news stories are interesting and not interesting.'
Nigel Dacre, ITN

A GOOD TV PROGRAMME HAS:

A good engaging story

Human interest, relating to viewers

A well-told narrative

Something new, or presented in a new way

Timeliness

Suitability for a particular, available slot in the schedules

You are going to create a programme on Brazil to be shown on *Newsround*. This can either be a presentation to the rest of the class (talk and poster) or a PowerPoint presentation.

Target task

1 Choose from the list of programme topics below:
- Life in a favela
- Sites and sounds of the rainforest
- 500 years of Brazilian achievements.

You should design a poster to illustrate the main ideas in your programme. You can use this to jog your memory when presenting. You should also draft out your talk, using key words for each idea.

Extension task

2 Choose from the list of programme topics below:
- 'If travelling is your passion, Brazil is your destiny.'
- 'All your dreams can come true in this land of happiness and hospitality.'
- 'São Paulo, the great investment opportunity in South America.'
- 'Sponsor a child today.'

Research the quote and create a programme, showing a balanced view of the issue.

Review

Activities

Word Association

1 For each word in the list write down the first thing that comes into your head about Brazil:

AMAZON

FOOD

FOREST

GEOGRAPHY

WILDLIFE

INTERNET

BRAZIL

AIRPORTS

SHIP

NIGHT

Compare your answers with those for the first word association you did earlier in the chapter (page 7).

What have you learnt about Brazil?

2 **a** Which Geography words and meanings have you learnt?
b What places can you name in Brazil?
c Can you remember any facts about Brazil (figures, percentages, etc.)?
d Describe how Brazil is linked to other countries.
e Compare Brazil with the UK. What is similar and what is different?

Maps of Brazil

3 Sketch this map of Brazil. Label it with up to 20 key facts, places or features that you can remember.

Answer to page 6, activity 1:
All the photographs are of Brazil.

▽ **Figure 1.35** Map of Brazil, showing its five main regions

Figure 1.35 Map of Brazil, showing its five main regions

2 Development: National Scale

Water – whose rights and responsibilities?

△ **Figure 2.1** Collecting water, UK
Scott Sinclair & Development Education Centre, Birmingham

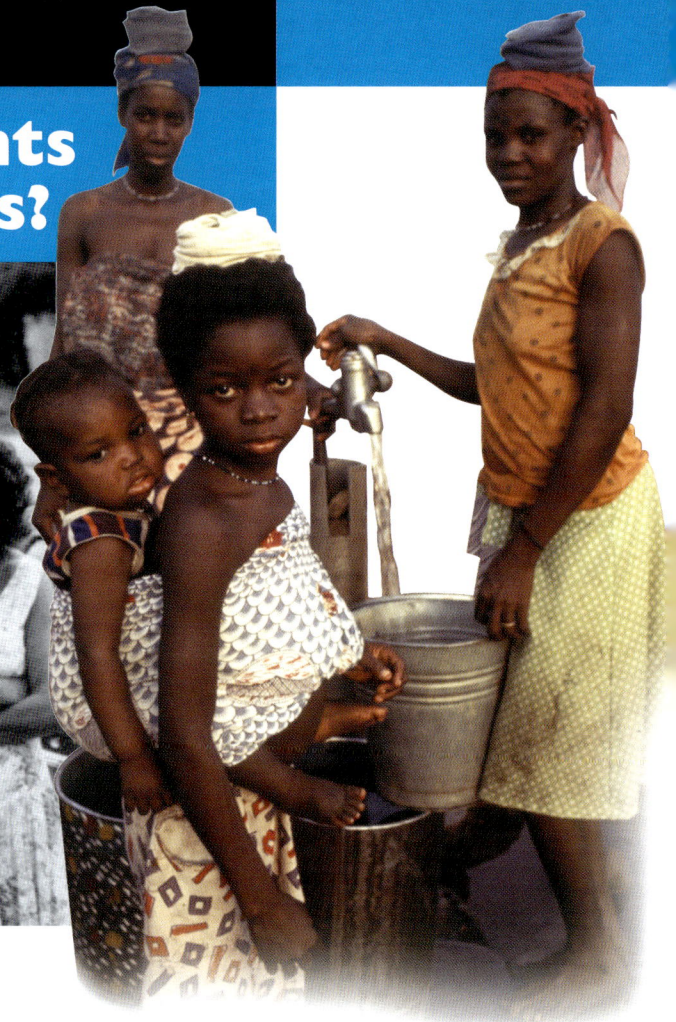

△ **Figure 2.2** Collecting water, Ghana

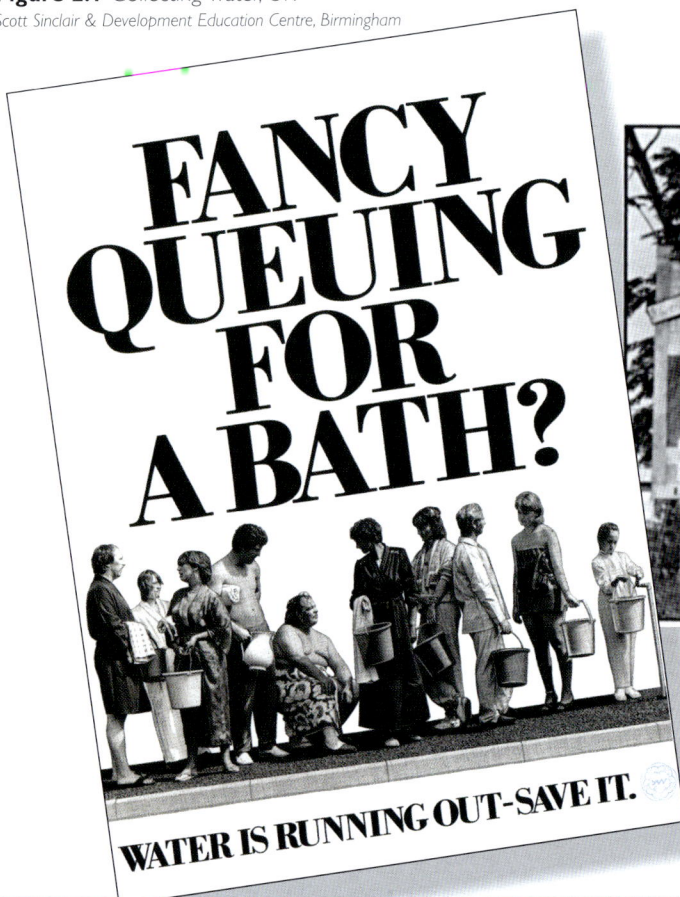

FANCY QUEUING FOR A BATH?

WATER IS RUNNING OUT – SAVE IT.

◁ **Figure 2.3** UK drought poster, 1976
Scott Sinclair & Development Education Centre, Birmingham

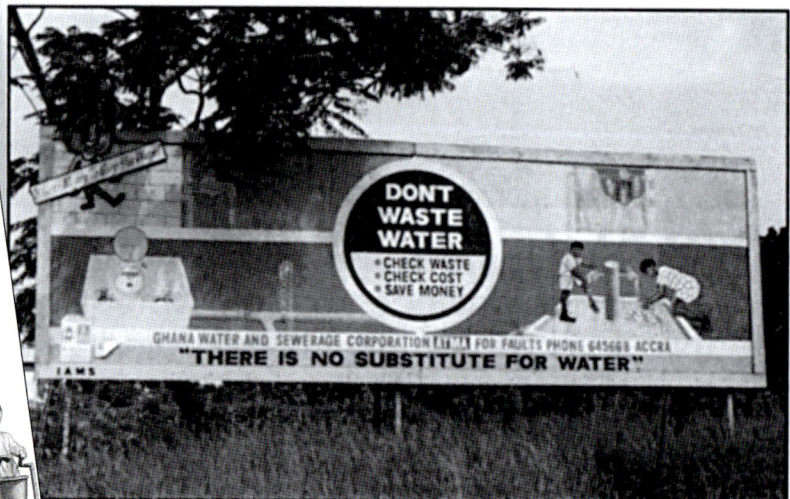

DON'T WASTE WATER
• CHECK WASTE
• CHECK COST
• SAVE MONEY

GHANA WATER AND SEWERAGE CORPORATION FOR FAULTS PHONE 645668 ACCRA
"THERE IS NO SUBSTITUTE FOR WATER"

△ **Figure 2.4** Ghana poster promoting careful use of water

Activities

1 These images are about access to water in Ghana and the United Kingdom. Use the frame opposite to help you think about these four photos. See how your frame is similar to and/or different from someone else's.

2 You have **rights** and **responsibilities** about using the **resources** on earth. What rights and responsibilities do you think that you have to using water? Write a list and keep it safe because you will need to use it later.

3 In countries where people have to carry their water home from a standpipe or well, the average **consumption** is 12 litres per person, per day. Fill in a copy of the water diary (Figure 2.6) for yourself for one day. What is your total? How does your consumption compare with 12 litres?

4 Do you now want to add or change anything on your rights and responsibilities list? You still need to keep it safe!

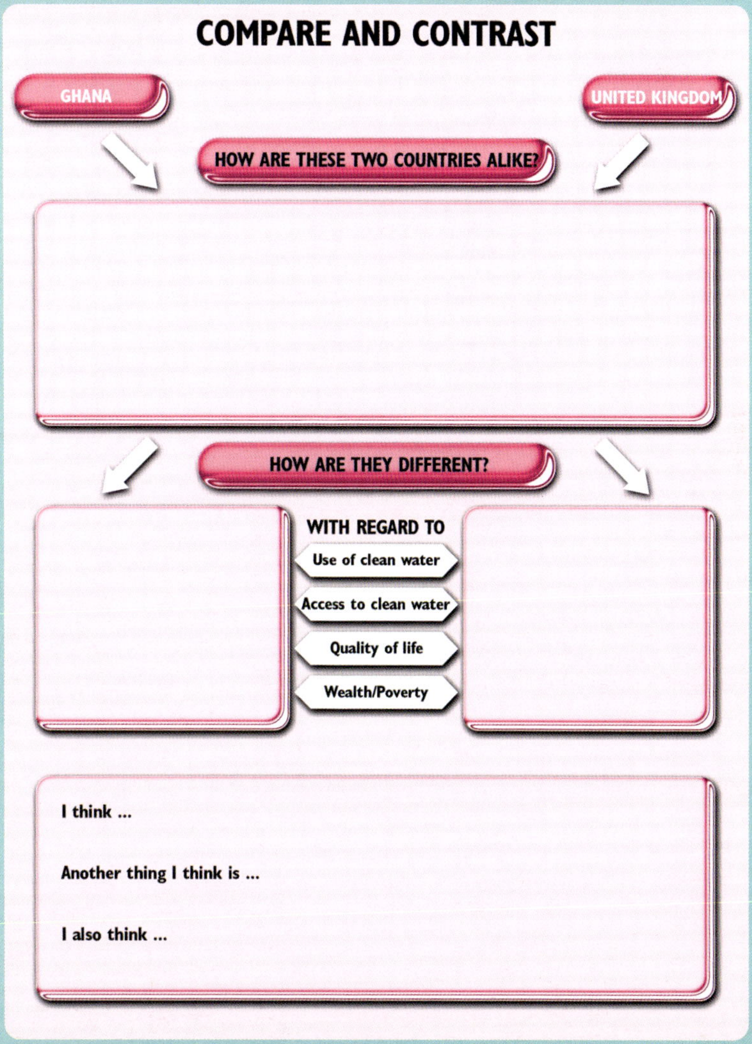

COMPARE AND CONTRAST

GHANA

UNITED KINGDOM

HOW ARE THESE TWO COUNTRIES ALIKE?

HOW ARE THEY DIFFERENT?

WITH REGARD TO
- Use of clean water
- Access to clean water
- Quality of life
- Wealth/Poverty

I think ...

Another thing I think is ...

I also think ...

△ **Figure 2.5** Compare and contrast frame

ACTIVITY	AMOUNT USED	NUMBER OF TIMES A DAY	TOTAL
Flushing the toilet	10 litres		
One bath	80 litres		
One shower	35 litres		
One washing machine	80 litres		
Cooking/drinking per person per day	10 litres		
Using a dishwasher	35 litres		

▷ **Figure 2.6** Water diary

Water – what is the reality?

Billions without clean water

Half the world's population is living in **unsanitary** conditions without access to clean water, according to a United Nations report published on Monday. The World Health Organisation says three billion of the world's most deprived people live in squalor and misery without access to proper **sanitation**. One billion of them have no access to safe water at all. But the UN says that this does not have to be the case. In its report, the UN argues that everyone could have clean drinking water and improved sanitation facilities within 25 years if **governments** made water provision a priority. It says access to water should be seen as a basic human **right** as well as a key factor in the fight against diseases such as typhoid and cholera.

UN water experts say 5 000 children die needlessly from water-borne illnesses.

The UN is calling on governments to concentrate on **community-based initiatives**, which it says are more cost effective and efficient than high tech centralised water supply policies.

Such projects in India, Bolivia, Ethiopia and Tanzania have dramatically improved people's living conditions and health levels, it argues. But, the UN warns, time is running out. Efforts to improve **global** hygiene are not keeping pace with the population explosion. If governments do not radically rethink their policies, the UN says, the world's water crisis will get worse.

△ **Figure 2.7** 'Billions without clean water' by Claire Doole in Geneva, Monday 13 March 2000

ICT links

If you do not already know about the United Nations and the World Health Organisation, the following websites might help:

www.unhchr.ch/

www.hrw.org/home.html

www.umn.edu/humanrts/treaties.htm

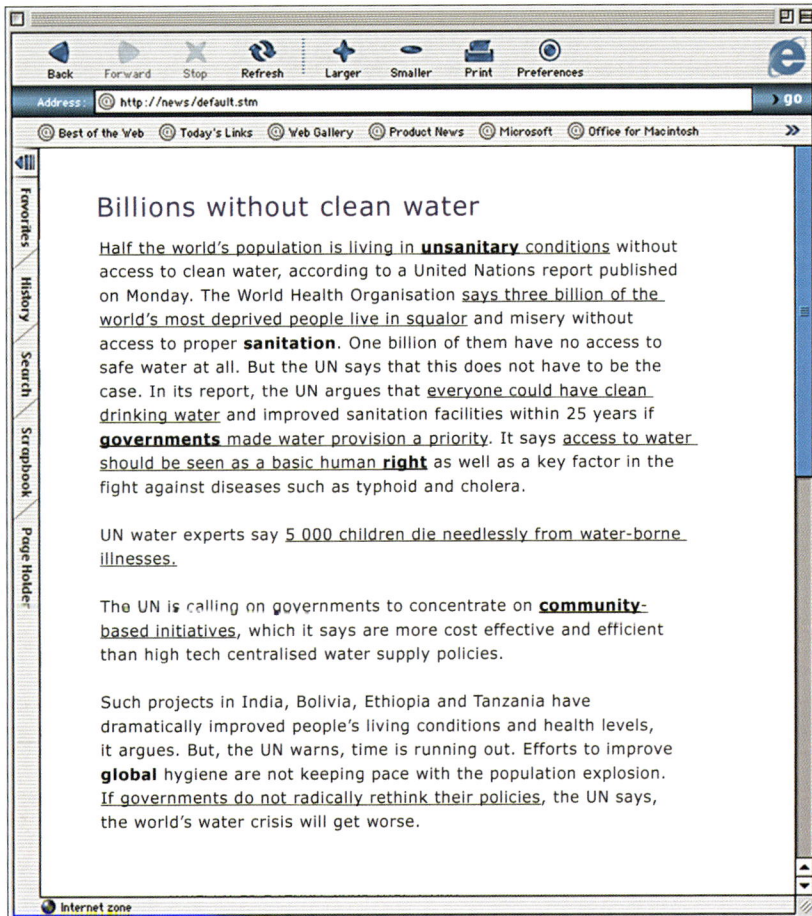

▷ **Figure 2.8** 5 000 children a day die from water-borne illnesses. Equivalent to 12 full jumbo jets

Activities

1 Read the article, 'Billions without clean water'. Now try to select the facts from the article. Put these in a table like the one opposite. A fact is something which is true …

FACTS	SOURCE
e.g 50% of the world's population does not have access to clean water	UN Health Report March 2002

Water – whose rights?

On pages 24 and 25 you have the opportunity to think about your rights to water. This is the **personal scale**. The article opposite (Figure 2.7) is also concerned with an individual's right to clean water. It challenges us with some horrifying images of children dying when their right to clean water is not respected. The article talks about the United Nations. This is an **international** organisation that was formed in 1944. It attempts to draw up worldwide rights. Not every country on earth is a member of the UN, but it is an organisation which has a global effect. Towards the end of the article, the writer states, 'If governments do not radically rethink their policies …'. This is a challenge to change at a **national scale**. It is mainly at this scale that we are going to investigate **development** issues relating to water.

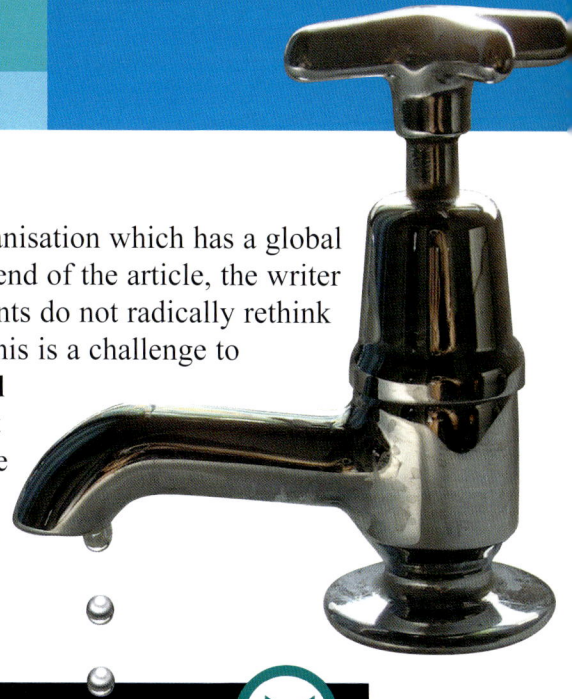

Activities

2 Read the article in Figure 2.7 again. Look carefully at the eight statements that are underlined. For each statement use the table below (Figure 2.9) to help identify the scale for which each statement is true.

GEOGRAPHICAL SCALE	DEFINITION	STATEMENTS WHICH ARE TRUE FOR THIS SCALE ARE:
PERSONAL	The space that surrounds you. A stranger in your personal space may make you feel uncomfortable.	
LOCAL	An area about the size of the catchment area of a secondary school. This area could be anywhere in the world, not just your home area.	
REGIONAL	This could be part of a country, e.g. the South East of England, or part of several countries, e.g. the Alps.	
NATIONAL	A whole political unit, e.g. the United Kingdom, or a country, e.g. France.	
INTERNATIONAL	This scale refers to links between two or more countries, e.g. trade.	
GLOBAL	Worldwide or whole world distributions, e.g. earthquakes, population densities.	

△ **Figure 2.9** Scale table

Water – whose reponsibility?

On these pages we are going to consider development related to water in Ghana and in the United Kingdom. The United Kingdom is made up of the countries of England, Scotland, Wales and Northern Ireland. However, it is the United Kingdom that these countries form that is recognised as one **nation** by the United Nations and many others. Most of the information will be at the national scale. It is difficult for all of our work to be at the national scale, since all scales are linked.

A nation can be said to be a group of people who share a common culture, history or language and have a feeling of national unity. A **state** is an area of land whose people have an independent government. A **nation-state** exists if a nation and a state have the same boundaries. An example of a nation-state is France. The term **country** refers to the land that a state occupies. These states have the responsibility of providing water for their people to use.

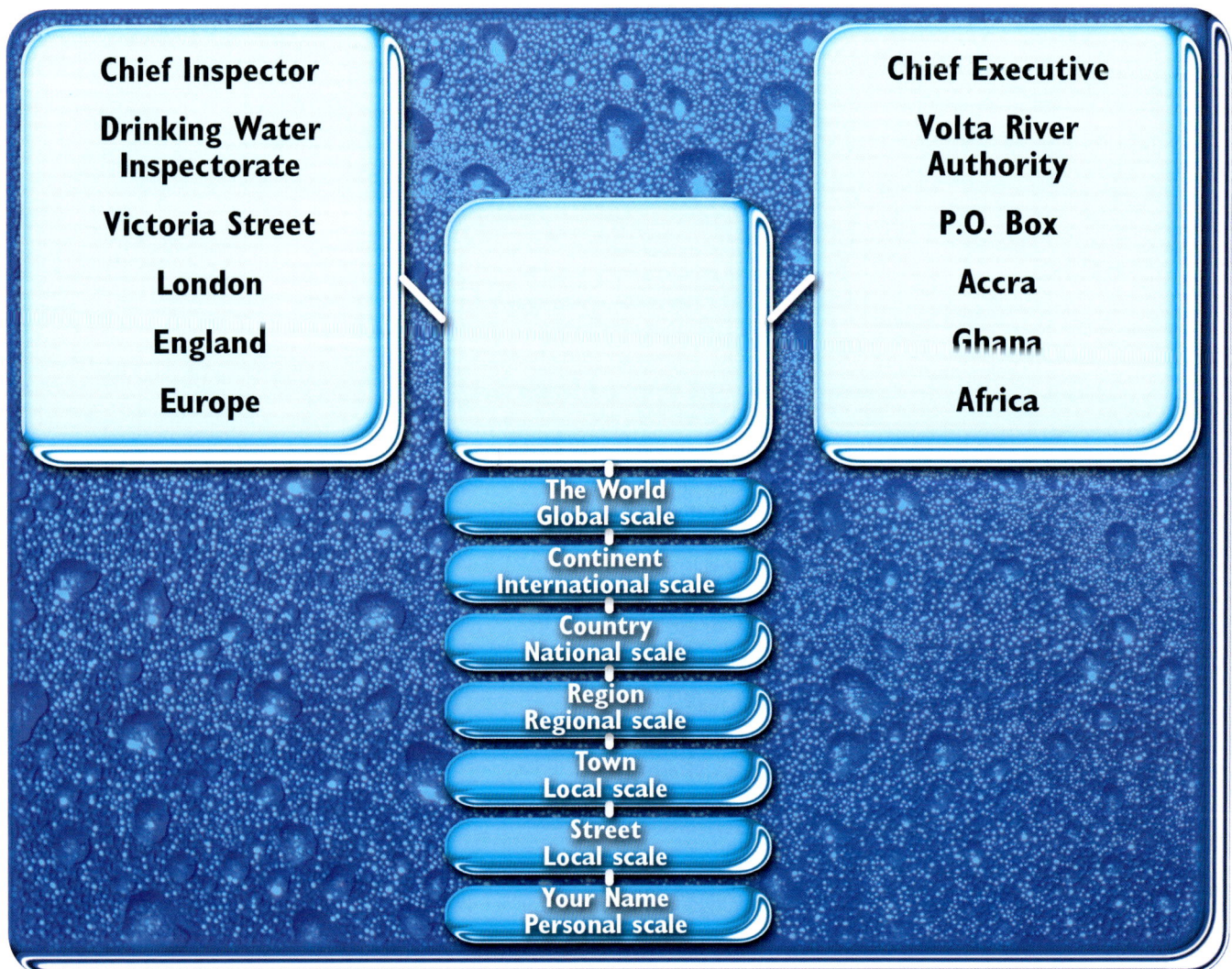

Chief Inspector

Drinking Water Inspectorate

Victoria Street

London

England

Europe

Chief Executive

Volta River Authority

P.O. Box

Accra

Ghana

Africa

The World
Global scale

Continent
International scale

Country
National scale

Region
Regional scale

Town
Local scale

Street
Local scale

Your Name
Personal scale

△ **Figure 2.10** Where in the world?

At which scale?

Activities

1 Think for a moment about your home address. Write it out, like the ones in Figure 2.10, and annotate it to show how you are linked to different places at different scales. The address is a universal way of placing us all on the earth. This is true for both people and places. The place where you learn has an address. It too links with many places at different scales.

2 Spend a moment thinking about yourself, which *nation*, *state* and *country* do you feel that you belong to? The three may be the same or different. Either is fine, there is no right answer, your views will depend on your own experiences. In fact you may feel that you are a member of two or more nations, e.g. American-Irish.

ICT activities

1 Use a search engine on the Internet. Search engines include:

Alta Vista   **www.altavista.com**

Ask Jeeves!   **www.ask.co.uk**

Exite   **www.excite.co.uk**

Yahoo   **www.yahoo.com**

Type in the key words: *water*, *development*, *national*. List the first ten addresses that you are shown. Why is www a term that links to the *global* scale?

2 How do you know from the Internet address which *country* the information comes from? How do you know whether the information comes from an organisation, or an individual? Why might these two pieces of information matter?

3 Working as a group of three, from your list of ten websites, choose one website to investigate further. Produce a summary of your website to share with others. To do this you need to:

a Produce a global address for the organisation or individual that has produced your website, like the ones on page 28.

b Provide a fact and source guide like the one on page 26. If there are also opinions and fiction on your website you might like to create new tables.

c Use the scale table on page 27 to help you to explore at which scale(s) information is given.

d Is there anything else that you would like to share about your website?

Some websites that your search engine might reveal are:

American Water Works Association
www.awwa.org/

Foundation for Water Research
www.fwr.org

United Kingdom Water Industry Research Ltd
www.ukwir.co.uk

Water Aid
www.oneworld.org/wateraid/

What is a nation?

This table presents a variety of information at the national scale.

FEATURES	GHANA	UNITED KINGDOM
Full Title	Republic of Ghana	United Kingdom of Great Britain and Northern Ireland
Flags		
Brief Description	A West African state bordering on the Gulf of Guinea, Ghana is bounded by Côte d'Ivoire to the west, Burkino Faso to the north, Togo to the east, and the Atlantic Ocean to the south.	The United Kingdom consisting of England, Wales, Scotland and Northern Ireland.
Government	President	Sovereign
Area	238 540 sq km	244 820 sq km
Maps	 Ghana	 United Kingdom
Population 2001 est.	19 894 014	59 647 790
Population Density	216 per sq mile	631 per sq mile
Infant Mortality	56.5/1 000	5.5/1 000
Average annual rate of natural population increase	1.9%	0.1%
Languages	English (official); Native tongues (Brong, Ahafo, Twi, Fanti, Ga, Ewe, Dagbam)	English, Welsh, Scots, Gaelic
Ethnicity/Race	Black African 99.8% (Major tribes: Akan 44%, Moshi-Dagomba 16%, Ewe 13%, Ga 8%), European and other 0.2%	English 81.5%; Scottish 9.6%; Irish 2.4%; Welsh 1.9%; Ulster 1.8%; West Indian, Indian, Pakistani and others 2.8%
Literacy Rates	60% (1990)	99% (1999)
Unemployment	20% (1997 est.)	6% (1999)
Labour Forces	4 million (1997) Agriculture 60% Industry 15% Services 25%	29.2 million (1999) Agriculture 1.1% Industry 18.7% Services 80.2%

▷ **Figure 2.11**
Data at the national scale

Activities

Foundation

1 Use three headings, *Nation*, *State* and *Country*. For each heading write a definition. Use page 28 to help. You may also wish to use the glossary on pages 124–29 and a dictionary.

2 Compare your definitions with those of two other people. Is there anything that you want to add or change about your definitions?

3 Working with a partner you need to choose the information from Figure 2.11 about either the UK or Ghana. Your teacher may tell you which to use. You now need to look at the information carefully and decide which heading each piece of information belongs to. For example, does the information about area tell you about the *nation*, *state* or *country*?

4 When you have put all the information under each heading you need to work with a new partner. Compare your lists with theirs. What information is the same and which is different? Together write a paragraph to explain why your lists were different. Can you produce a new improved definition for each heading?

Target

5 Draw an outline map of the United Kingdom.

6 Re-read page 28. Using this and the glossary on pages 124–29 explain what you think the difference is between a *nation*, a *state* and a *country*.

7 Annotate your map using the information from page 30. Use one colour for information that tells us more about the *nation*, another for information that tells us about the *state*, another for information that tells us about the *country*. Add a key.

8 Compare your annotated map with those of two other people. Are there any changes that you wish to make? Write a paragraph to explain any changes.

9 Now do the same for Ghana.

10 Are either the United Kingdom or Ghana nation-states? Write a paragraph about each place using your annotated maps to help you explain your answer.

Extension

11 Using the information about a *nation*, *state* and *country* on page 28, and the information about the United Kingdom and Ghana on page 30 to help you, complete the table below. Make your copy large enough so that you can write a paragraph in each cell.

	This is a NATION because ...	This is a STATE because ...	This is a COUNTRY because ...
GHANA			
UNITED KINGDOM			

12 Why can geographers' use of the term 'national scale' be confusing? Can you think of anything better?

13 In Jamaica, 'out of many – one people' is a national motto. Could this statement be important for Ghana? For the United Kingdom? For both places? Using the information from your table to illustrate your points, explain your views. You need to back up each idea with information.

Different measures, different realities?

When working at the national scale, development has often been seen to be the same as economic growth. Whilst it is useful to compare wealth between countries, it tells us nothing about how wealth is divided up within countries, and it does not measure non-economic factors.

So how else can we measure development? There are a number of different strategies. On these pages we are going to choose two different methods. One is to use the Human Development Index (HDI). The other is to take aspects of the UN Declaration of Human Rights, and see how the UK and Ghana as two nations measure up in terms of water-related developments.

Everyone has the right to life, freedom and to security.

Everyone has the right to a nationality.

Everyone has the right to a standard of living adequate for health and well-being.

Everyone has the right to adequate food, clothing and shelter.

Everyone has the right to own property either alone or with others.

Everyone has the right to education.

Everyone has the right to meet together peacefully.

Everyone has the right to a world system, which guarantees all the rights of this declaration.

△ **Figure 2.12** Extracts from the UN Declaration of Human Rights

Activities

1 The rights in Figure 2.12 are taken from the UN Declaration of Human Rights. Rank them in your order of importance.

2 Having done this, working in a group of three to four people, try to agree on two or three rights that you all think are the most important. Share this with the class.

3 Which came out overall as the most important?

4 Why do you think this was? Was it difficult or easy to agree?

The Human Development Index

The data on page 33 comes from the 1999 Human Development Report. The Human Development Index (HDI) measures the progress of a country in terms of human development. The world map (Figure 2.13) shows the 174 countries that are ranked and grouped into three levels of development: low, medium and high. Table A shows Ghana, ranked 129th, and the three countries above and one below. Table B shows the United Kingdom, ranked 10th, and the three countries above and one below. The data relates to access to sanitation. This is a key feature relating to water and health and has a link with many of the rights illustrated in Figure 2.12. Services such as sanitation are the responsibility of nations supporting the rights of the individual. Sanitation means taking toilet and other waste water and turning it into products that can be re-used, e.g. clean water and fertilisers.

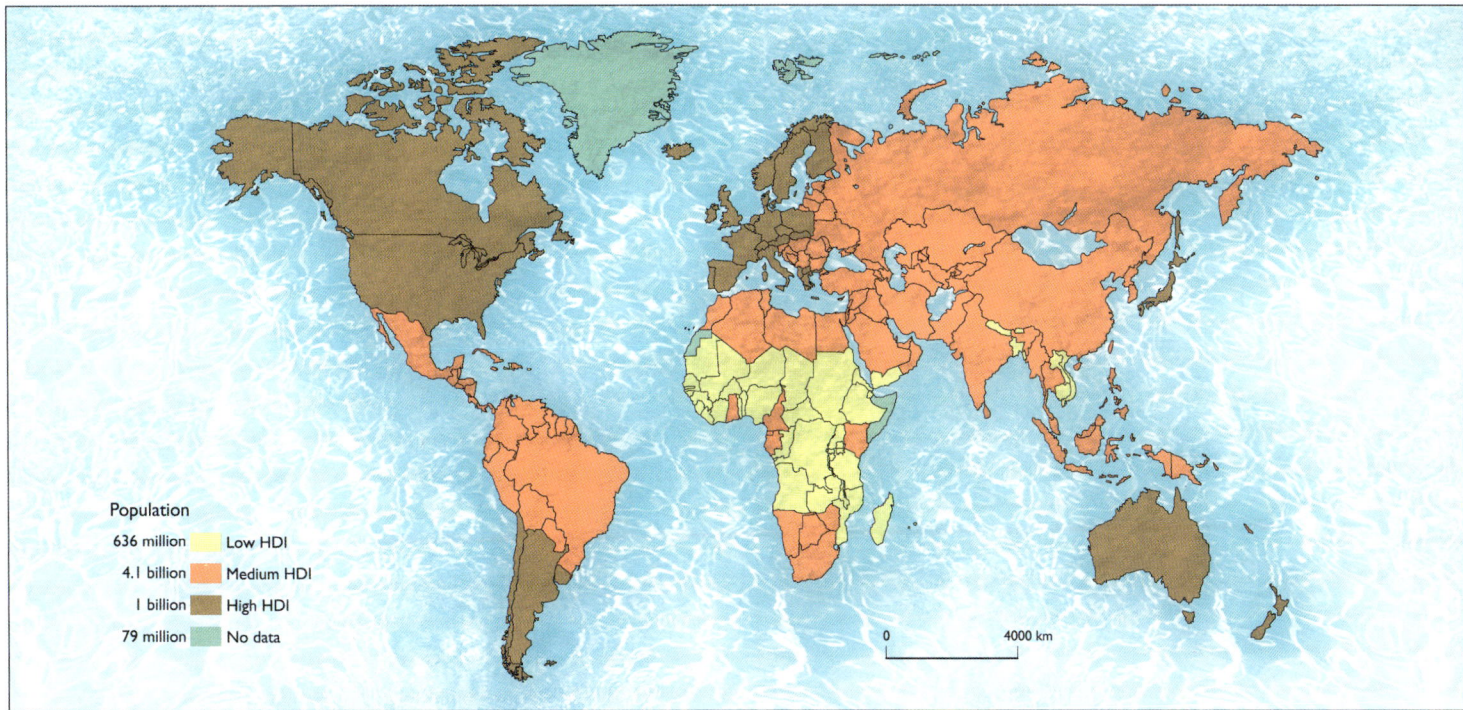

△ **Figure 2.13** Human Development Index map, 1999

Population
636 million — Low HDI
4.1 billion — Medium HDI
1 billion — High HDI
79 million — No data

0   4000 km

HDI Ranking	Country	% of population served by public sanitation (1997)
126	Iraq	75
127	Lesotho	38
128	India	29
129	Ghana	32
130	Zimbabwe	52

△ **Figure 2.14** Table A – sanitation information

HDI Ranking	Country	% of population served by public sanitation (1997)
7	Belgium	75
8	Netherlands	98
9	Japan	55
10	U.K.	96
11	Finland	77

△ **Figure 2.15** Table B – sanitation information

Activities

5 On a copy of a world map, mark on the outline and label the ten countries listed in Tables A and B. Label the seven continents on your map. You can use an atlas to help you.

6 Use three colours to produce a key to show low, medium and high human development. Now use the map in Figure 2.13 to colour in the ten countries on your world map.

7 Look at the list of rights on page 32. Which rights are improved by a nation with good sanitation?

8 Why do you think that it might be easier for some nations rather than others to improve access to sanitation? Try to give examples from both Tables A and B to support your answer.

9 Re-read the newspaper article on page 26. What type of development project would you suggest for a nation hoping to improve its sanitation services?

Ghana: water and development – a reality?

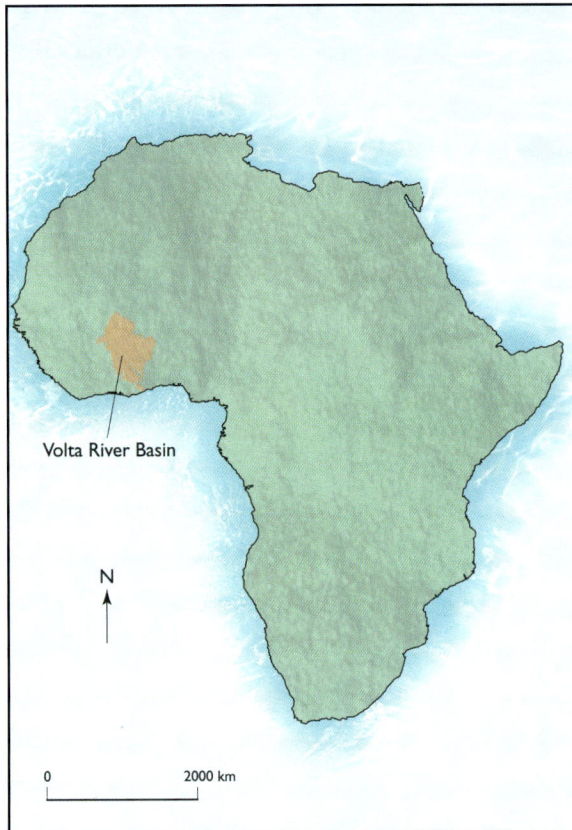

Volta River Basin

N

0   2000 km

FACT FILE: THE VOLTA BASIN

The Volta River is the main fresh water source for Ghana.

The Volta River is formed by the confluence (the place where two or more rivers join together to form one) of the Black Volta and the White Volta rivers at Yeji in the central part of the country.

The river flows in a southerly course through Lake Volta to Ada on the Gulf of Guinea.

The total length of the river, including the Black Volta, is 1 500 km.

Lake Volta was created by the construction of the Akosombo dam on the river.

At about 8 482 sq km, the lake is one of the largest artificially created lakes in the world.

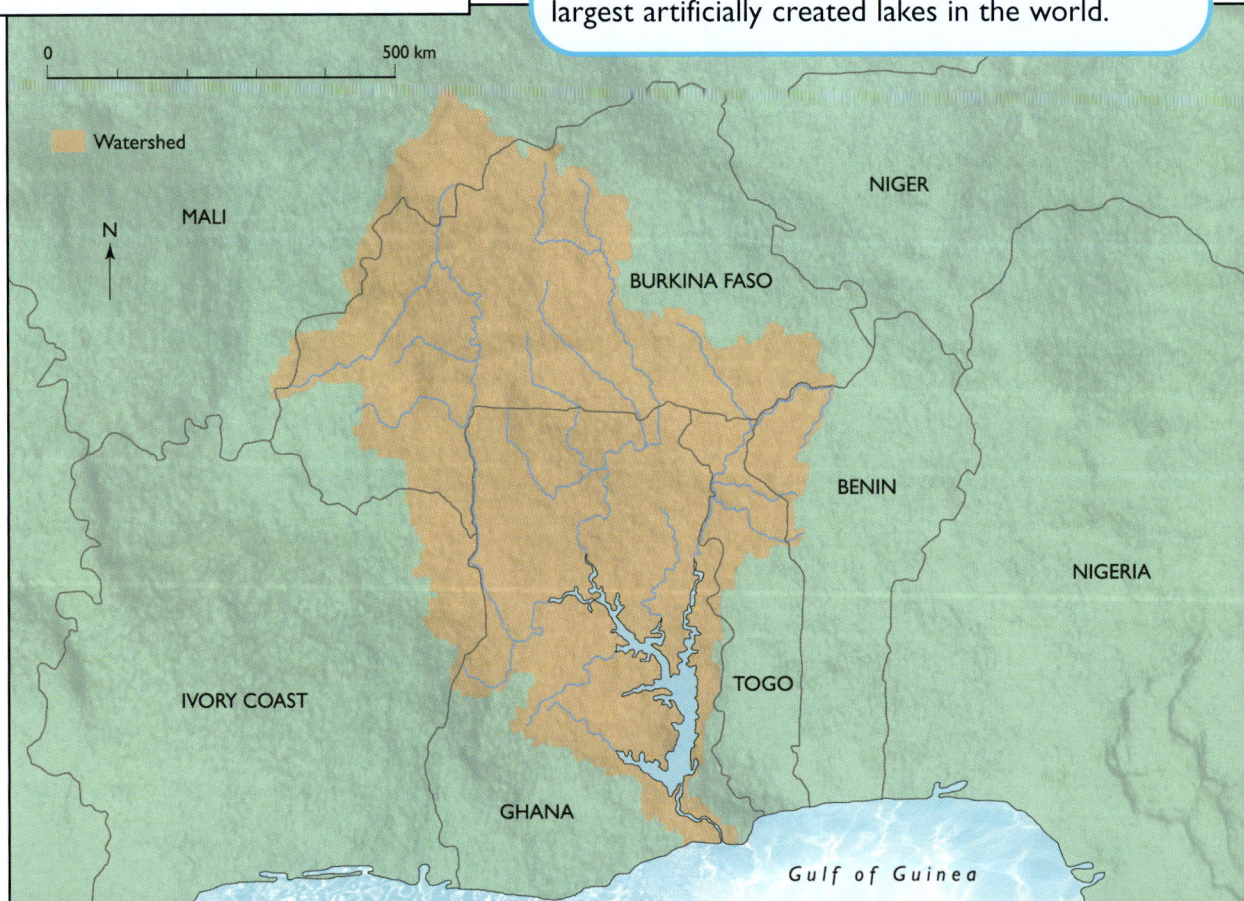

0   500 km

Watershed

N

MALI

NIGER

BURKINA FASO

BENIN

NIGERIA

IVORY COAST

TOGO

GHANA

Gulf of Guinea

△ **Figure 2.16**
Africa and the Volta basin

△ **Figure 2.17** Boys fishing in the Volta River

△ **Figure 2.18** A satellite view of the Volta River

△ **Figure 2.19** Volta River and the Akosombo dam

The Volta Project – different realities?

The Akosombo Lake is an artificial lake created after the River Volta was damned at the Akosombo gorge. The lake was created to store up water mainly to generate hydroelectric power (HEP). It was also hoped that by creating the lake, inland water transport would be improved, there would be greater opportunities for the local fishing industry, and there would be enough water for domestic and industrial use and for irrigation. The project cost £70 million and was completed in 1966. It was a national project, funded in part by the World Bank and the USA and UK.

National project, local impacts

The local fishing industry now supplies 10% of the country's fish. The lake has supported some small-scale irrigation projects to improve farming. When it was built the lake submerged nearly 740 villages and displaced 80 000 people. About 70 000 were moved to newly-built settlements, which were supposed to have small concrete houses and services, such as schools and wells, and agricultural machinery. Many of the settlements had problems and the majority of people settled in other places. Cases of water-related diseases such as malaria increased near the river. The lake has caused problems for trade between north-eastern and southern Ghana. In 1997, 70% of the population was still without electricity.

› Droughts cause energy crisis

In 1999 the Volta River Authority in Accra produced a poster that said 'Save Akasombo lake, reduce energy consumption'. This was a reaction to the 1990s' droughts and very low water levels in the lake. The low water levels meant that not so much electricity could be produced, and that energy exports were stopped. Now Ghana is exploring other energy supplies, for example gas and thermal energy.

△ **Figure 2.20** Pros and cons diagram

Activities

Foundation

1 Working with a partner, look carefully at the information on pages 34, 35, and 36. On a copy of the pros and cons diagram (Figure 2.20), fill in the bubbles with as many advantages of the Volta project as you can.

2 Look at page 32 again and the list of Human Rights. Which rights do you think the government hoped to improve by building the Akasombo dam?

Target

3 Make a copy of the pros and cons diagram, look again at pages 34, 35 and 36, and fill in the disadvantages.

4 Look again at page 32. Which human rights were not being respected?

5 Why did the Akasombo dam not work as planned?

Extension

6 Look again at the article on page 26 (Figure 2.7). Can you suggest a different style of development project that may have had greater success?

7 There is a Native American saying, 'Only when you have felled the last tree, caught the last fish and polluted the last river will you realise that you can't eat money.' Using the building of the Akasombo dam as your case study, explain what you think that this saying means.

The United Kingdom: clean drinking water – a reality?

The United Kingdom's water industry:

- provides 18 000 million litres of water every day to 58 million people – that's enough to provide everyone on earth with twice the recommended intake

- **spends around £3 billion every year on improving the water supply and sewerage services**

- has 397 401 km of water mains – and 354 066 km of sewers – enough to stretch to the moon and back

- **has around 2 250 water treatment works and 9 260 sewage treatment works**

- has 1 584 boreholes, 666 reservoirs, and 602 river abstractions – two thirds of our water comes from surface water and one third from groundwater

- **carried out over 2.8 million tests to check water quality. 99.82% of samples met the standards. Only 3% of drinking water is used for drinking and cooking – washing and toilet flushing use large amounts**

- the average domestic bill for water and sewerage services is 64p per day – one glass of water costs 0.03p, a bath 10p, and a shower 4p

- **provides each person in the UK with 343 litres per day**

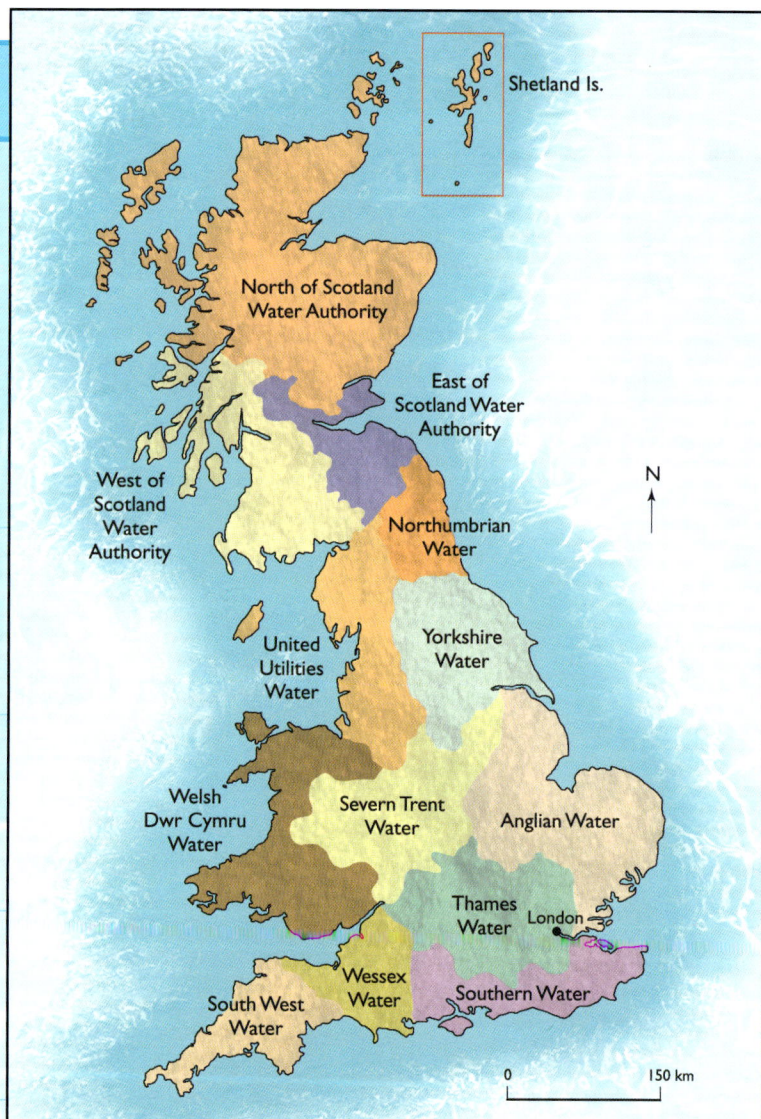

△ **Figure 2.21** Map of the UK's water boards

△ **Figure 2.22** Water facts, UK 2002

△ **Figure 2.23** Where is this?

Figure 2.24 shows the poster promoting the Water Matters Campaign. This is a temporary campaign that has been initiated by WaterAid and Tearfund. It is a campaign that wants to make world politicians more aware of the global water crises. They are concerned that the UN predicts that two out of three people will be living with water shortages by 2025. Living in the UK it can be hard to believe that the world is facing a water crisis. Water Matters believes that over the next 10 years, 20 million children will die of preventable water-related diseases.

make a splash FOR THE WORLD'S POOR

Water Matters

△ **Figure 2.24** Poster promoting the Water Matters Campaign

Activities

1 Look at your estimate for your own personal water use from page 25. How does it compare with the average 343 litres used per day for the UK? Why do you think that your own usage is higher or lower than this average?

2 Look at the map and water facts opposite. Using the list of human rights on page 32, for each fact decide which rights are being supported at a national level.

3 The answer to the 'Where is this?' question at the top of this page is Birmingham, in 1843. Birmingham would be described differently now. As a place, it shows that change is possible. Using the information on this page, describe for Water Matters what you as a global citizen can do to help the world's water crisis. Try to include actions as well as thoughts in your description.

› What path to development?

Some geographers used to think that each country, no matter when or where, had to pass through different stages on the path to development. Some would argue that Birmingham in 1843 could be similar to some less economically developed countries today. The activities on pages 40–42 will show that development is far more complex than this.

Assessment tasks

UK – clean drinking water for all?

Private water supplies 'E.coli risk'

Four out of five private water supplies in the UK fail to meet EU drinking standards and put more than half a million people at risk of E.coli, a BBC investigation has found.

A leading microbiologist has described the situation as a 'mess', and called for a tightening of the current regulations on water quality testing.

Water testing

In Britain there are at least 300 000 private water supplies.

The majority are in the countryside and they can range from a well supplying a single farm to large businesses. Some even supply hospitals.

They are defined as water that is not provided by a licensed water undertaker, in other words, a water company.

The extent of water testing appears to rely upon the number of people using it.

The regulations state that if a private supply provides water for one house then it may never be tested – and even if there is an adjacent campsite or caravan park it is only tested once a year.

If it supplies 500 people – and that could be a small village – then it need only be tested twice a year.

△ **Figure 2.25** Water supplies at risk?

Assessment tasks

Development – at what scale?

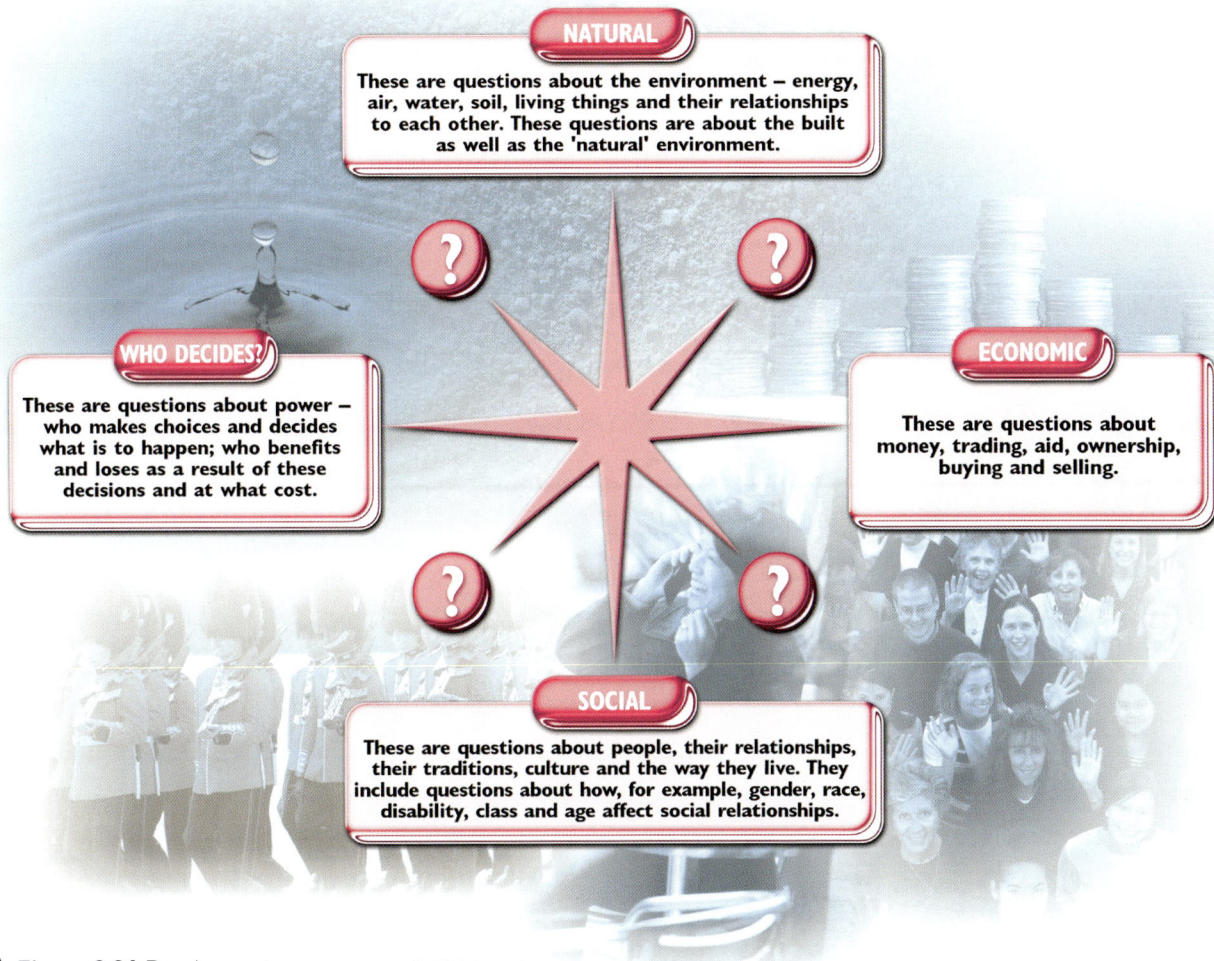

△ **Figure 2.26** Development compass rose © *DEC, Birmingham*

Target tasks

1 On a blank copy of the development compass rose, using the facts and figures on page 38, and any other information that you have about the UK water industry, decide which facts belong to each heading. All this information is at the national scale.

2 Read carefully a copy of the news article on page 40. On a second blank copy of the development compass rose, decide which pieces of information from the article belong under each of the headings.

3 Use both your development compass rose diagrams to explain how the private

water supplies are different from the water companies. You may wish to use the four compass rose headings to help you decide this.

4 Look again at the list of human rights on page 32. If you were supplied with polluted water, which of your human rights would not be supported?

5 Why is it so difficult to describe an entire nation or country as being less economically developed or more economically developed?

△ **Figure 2.27** Former Chief Executive of Volta River Authority, Mr E. A. K. Kalitisi

'When you travel around this country, you will find bamboo poles all over the place, it is like a forest of trees but these if you watch carefully are antennas for television. So what it means is that in the furthest parts of this country, wherever we have electricity, we have been able to link our rural population to the rest of the world.

So I believe that this is one of the most powerful means of elevating people from the poorer countries into the rest of the world. It has become more powerful than literal education which we used earlier to stimulate development in the rural areas. Now with television and radio electricity has enabled the countryside to have immediate access to the world.'

△ **Figure 2.28** Former Chief Executive of Volta River Authority, Mr Kalitisi, August 1997

Extension tasks

1 Using a blank copy of the development compass rose (Figure 2.26) from page 41, show the local impacts of the Volta River project. Use the information on page 38 to help you.

2 On another copy of the development compass rose, use the former Chief Executive's comments and the information from page 38 to help you to fill in the information for the Volta Project at the national scale.

3 Look carefully at your two completed development compass roses.

a Why is development such a complex theme to study in geography?

b Why does scale matter?

4 Read the former Chief Executive's views about development. He believes that the Volta Project had a greater impact on local development than education. Using evidence from your work in this chapter and anything else you know about development, explain why you might agree or disagree with Mr Kalitsi.
You might want to use these ideas:

I agree/disagree with the former Chief Executive because …
Another reason is …
I also think that …
I also agree /disagree that …
Finally I think that …

The right to develop?

Activities

1 Use again the compare and contrast frame (Figure 2.5) from page 25. Fill it in using what you now know about water issues in the UK and Ghana. How is this similar and how is it different to the first one that you completed?

2 Look back at your answer to question 2 on page 25. Is there anything that you might like to add?

3 Re-read the Internet article (Figure 2.7) on page 26, and look again at the Human Development Index map (Figure 2.13) on page 33.

a On a blank world map, using the low, medium and high categories, create a map of what you hope will be the state of development for the ten countries from page 33 in 25 years' time. Annotate the map to explain any changes that you would like to happen.

b Give examples of how the changes that you have suggested might affect human rights at a local, national and international scale.

c Construct a list of between five and ten actions that would need to happen to make your future world a reality.

d Look at these actions; who needs to take responsibility for these, and at what scale?

▽ **Figure 2.29** Water images from the UN

Coasts

What do we do beside the seaside?

▷ **Figure 3.1** Wide open sands and a Mediterranean climate attract people to Alicante, Spain

◁ **Figure 3.2** People watching migrating birds on the sandy coast of Northumberland

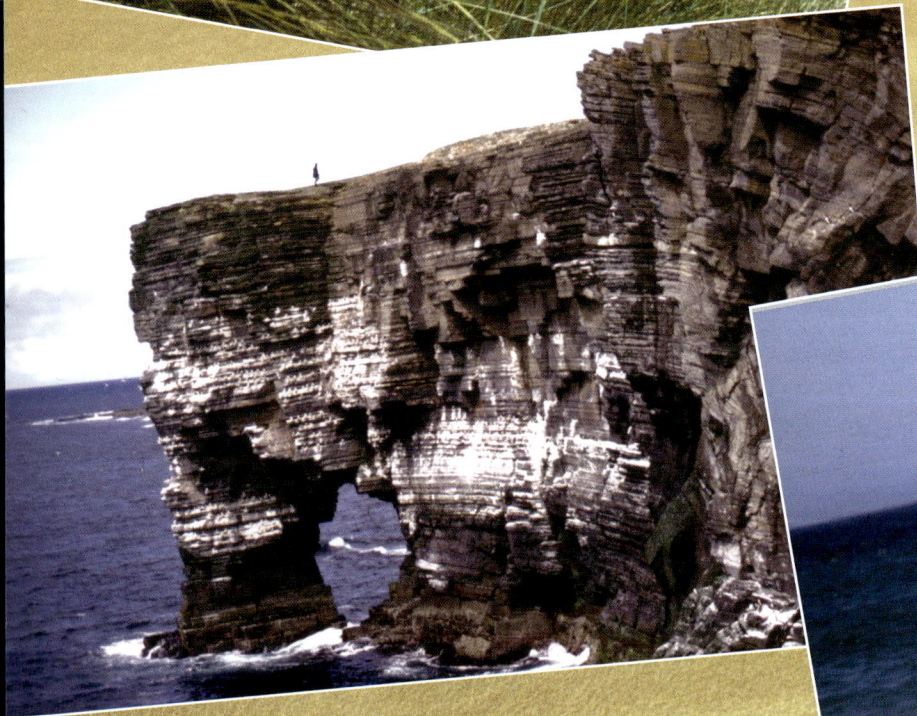

◁ **Figure 3.3** A single walker on the spectacular scenery of the cliffs of Ronsay, Orkney

▷ **Figure 3.4** Windy coasts and long distances out to sea create big waves for surfing, Cape Wookimai, Victoria, Australia

Oh I do like to be beside the seaside.
Oh I do like to be beside the sea.
Oh I do like to stroll along the prom prom prom,
Where the brass bands play tiddley om pom pom.
Oh I do like to be beside the seaside.
Oh I do like to be beside the sea.

△ **Figure 3.5** Traditional British seaside song from Victorian Era

Activities

1 Do you like to be beside the sea? If so where, when and why?

2 This old song (Figure 3.5) needs new words. Rewrite it with what you like to do. It may also need a new tune! How about a seaside rap?

3 Devise a questionnaire survey of 20 people. Include questions about:
- where people have been,
- what people do there,
- why people liked it or not.

Present your findings as both maps and graphs.

ICT activity

Present your findings as a series of graphs using **spreadsheet** software. Obtain a digital version of the map and use **graphics** software to add symbols to your map.

WARNING! There are events described on the next two pages that might upset you.

Why is it important to understand the coast?

On average, 500 people drown each year in British waters. This **pie chart** (Figure 3.6) shows the percentage of people who drown in different types of waters. What does the chart tell you about *where* people drown? How many drown in the sea? But sadly drowning is not the only accident that can happen at the coast. Read the articles opposite. They are true stories.

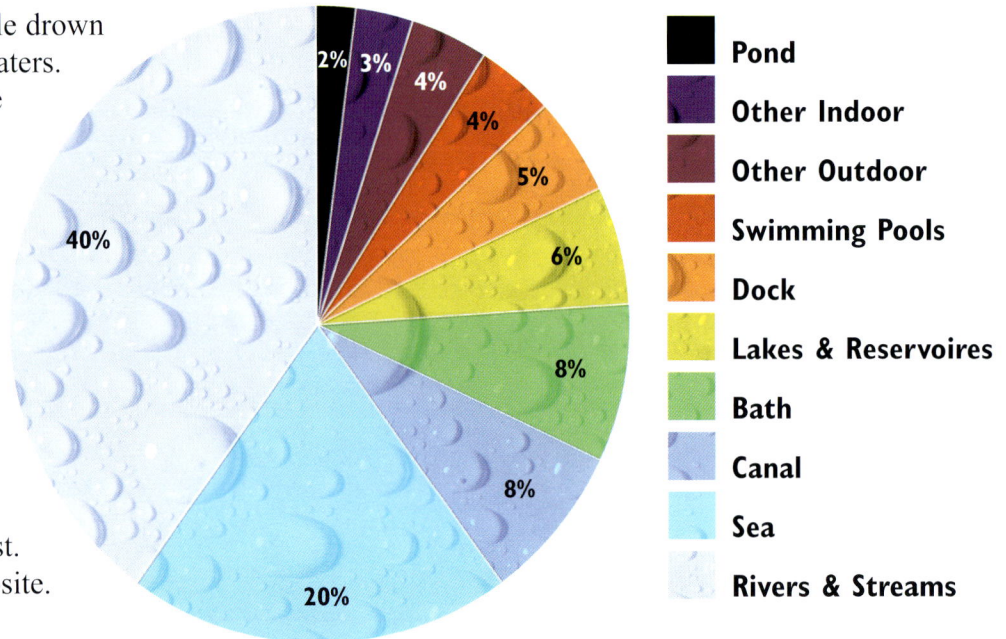

Pie chart values: 2%, 3%, 4%, 4%, 5%, 6%, 8%, 8%, 20%, 40%

Legend:
- Pond
- Other Indoor
- Other Outdoor
- Swimming Pools
- Dock
- Lakes & Reservoires
- Bath
- Canal
- Sea
- Rivers & Streams

△ **Figure 3.6** Deaths by drowning in the UK – types of water – source: The Lifesaving Society

▽ **Figure 3.8** Beachy Head, East Sussex, is almost vertical because it is rapidly eroding. It is approximately 150 metres high which is as high as 30 double decker buses.

△ **Figure 3.7** The Menai Straits between Anglesey and North Wales at low tide. Tides sweep up gullies and creeks leaving sandbanks as islands. These will be the last places to be covered by the sea.

PRESS RELEASE   PRESS RELEASE   PRESS RELEASE

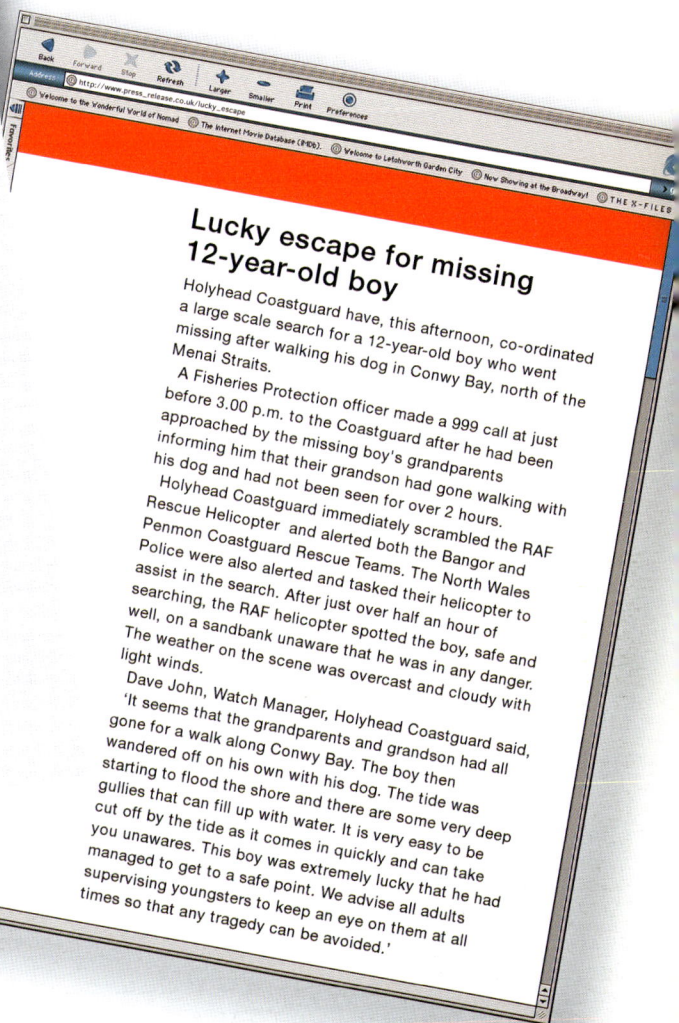

GIRL FALLS FROM CLIFFS AT BEACHY HEAD

At 10.15am this morning Dover Coastguard received a 999 call from a member of the public stating that a teenage girl had fallen from the cliffs at Beachy Head. It is believed that the girl had lost her footing, before falling 400 ft from the top of the cliffs to her death.

Dover Coastguard immediately tasked the helicopter from Lee-on-Solent to the scene as well as an inshore and an offshore lifeboat from Eastbourne. Eastbourne Cliff Team and Police were also tasked to the area. The first units arrived at the scene within 10 minutes and the casualty was located soon after at the base of the cliffs at Gun Gardens. She was recovered by the inshore lifeboat and was taken to the Lifeboat Station at Eastbourne. The Police dealt with the incident from this point onwards.

Phil Myers, Watch Manager said: 'This was a very tragic accident. We believe that the girl slipped on the wet grass and chalk when she was too close to the edge of the cliffs. The other teenagers within the party saw her falling from the cliffs and are now trying to come to terms with the accident. We urge people to be very careful when they are on the cliffs and particularly to stay well away from the edge.'

PRESS RELEASE   PRESS RELEASE   PRESS RELEASE

Lucky escape for missing 12-year-old boy

Holyhead Coastguard have, this afternoon, co-ordinated a large scale search for a 12-year-old boy who went missing after walking his dog in Conwy Bay, north of the Menai Straits.

A Fisheries Protection officer made a 999 call at just before 3.00 p.m. to the Coastguard after he had been approached by the missing boy's grandparents informing him that their grandson had gone walking with his dog and had not been seen for over 2 hours.

Holyhead Coastguard immediately scrambled the RAF Rescue Helicopter and alerted both the Bangor and Penmon Coastguard Rescue Teams. The North Wales Police were also alerted and tasked their helicopter to assist in the search. After just over half an hour of searching, the RAF helicopter spotted the boy, safe and well, on a sandbank unaware that he was in any danger. The weather on the scene was overcast and cloudy with light winds.

Dave John, Watch Manager, Holyhead Coastguard said, 'It seems that the grandparents and grandson had all gone for a walk along Conwy Bay. The boy then wandered off on his own with his dog. The tide was starting to flood the shore and there are some very deep gullies that can fill up with water. It is very easy to be cut off by the tide as it comes in quickly and can take you unawares. This boy was extremely lucky that he had managed to get to a safe point. We advise all adults supervising youngsters to keep an eye on them at all times so that any tragedy can be avoided.'

△ **Figure 3.9** Press releases on the Marine and Coastguard Agency website. Sadly there are more at: www.mcagency.org.uk

Activities

1 Please take a minute's silence to think of the grief and worry of the families of these young people and what they must have felt.

2 Discuss what we should be doing about such tragedies and near misses. What could we do, in our geography lessons, to help stop anyone from our community drowning near the sea or falling from cliffs?

3 What are the geographical questions that you need to ask to find out more about the coast? How might you order them into a geographical enquiry?

What are the processes shaping the coast?

Process 1 – Weathering is breakdown!

Weathering is happening everywhere including the coast. The house in Figure 3.10 is falling down. It is being attacked by the weather and has been left alone to rot. Rotting is really weathering. Weathering is the breakdown of rocks and materials in three ways:

Physical weathering is the breakdown by wetting and drying, freezing and thawing or heating and cooling.

Chemical weathering is the breakdown by chemical reactions such as corrosion (oxidation) and dissolving (solution).

Biological weathering is the breakdown by animals and plants, such as root growth and burrowing.

▽ **Figure 3.10** How a house is affected by weathering

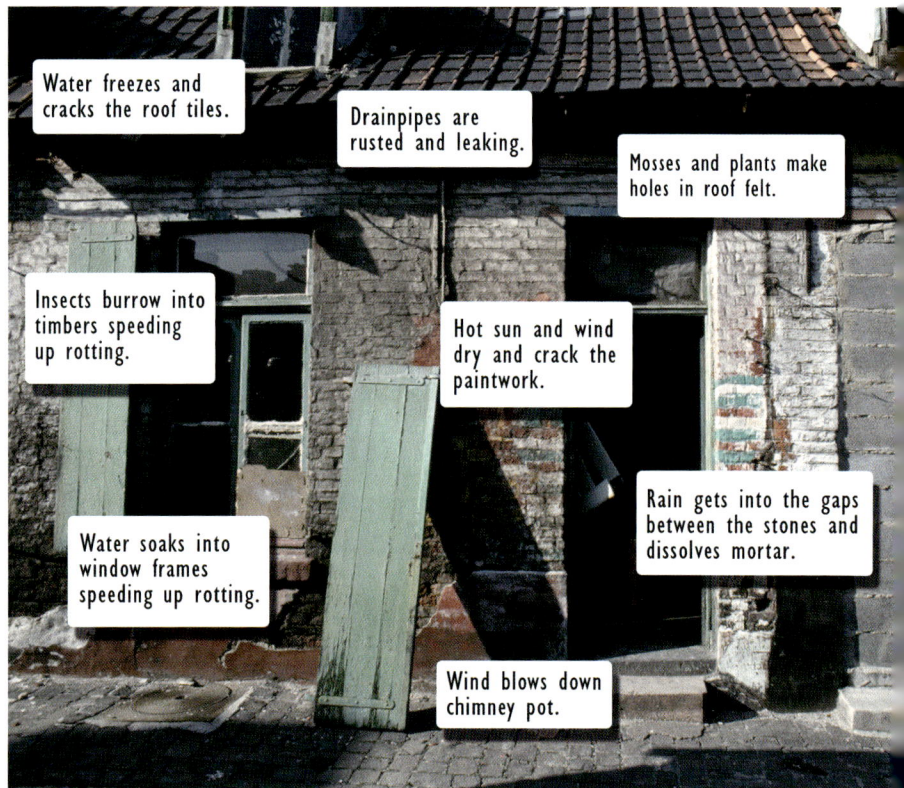

Water freezes and cracks the roof tiles.

Drainpipes are rusted and leaking.

Mosses and plants make holes in roof felt.

Insects burrow into timbers speeding up rotting.

Hot sun and wind dry and crack the paintwork.

Rain gets into the gaps between the stones and dissolves mortar.

Water soaks into window frames speeding up rotting.

Wind blows down chimney pot.

▽ **Figure 3.11** Weathering affects coastal cliffs too. Rocks very often get wet then dry then wet again. Wind is often much stronger. Salt spray reacts with some rocks, dissolving chemicals. Plants grow on ledges and seabirds burrow into cliffs. Durdle Door, South Dorset

Activities

1 Classify the annotations in the picture (Figure 3.10) into the three types of weathering.

2 Suggest ways of protecting a building against each weathering problem.

3 Using the photograph in Figure 3.11, draw a quick sketch and annotate how a cliff is affected by weathering.

4 Have you ever noticed that, near the sea in the UK, many houses have their outside walls painted? List three to five reasons why that might be.

Process 2 - Erosion is wearing away

The sea is one agent of **erosion**, wearing away the land. There is a bible story about the two men who built houses near the sea. One built on sand and the other built on rock. One lost his house and the other didn't! Why? The rest of the story is coastal erosion.

❯ Erosion at the coast means CASH

Erosion costs money, causes the loss of land, the loss of buildings and the loss of sea defences. The CASH acronym helps us remember the four processes of coastal erosion:

C is for **corrasion** (abrasion) – the stones and sand in the water grinding the rock away.

A is for **attrition** – particles in the water wearing away and rounding each other.

S is for **solution** – seawater dissolves chemicals in the rocks.

H is for **hydraulic action** – air pockets trapped in the waves create pressure and can split rocks apart.

◁ **Figure 3.12** At Flamborough Head, North Yorkshire, the **chalk** cliffs are mainly eroded at high tide in stormy weather

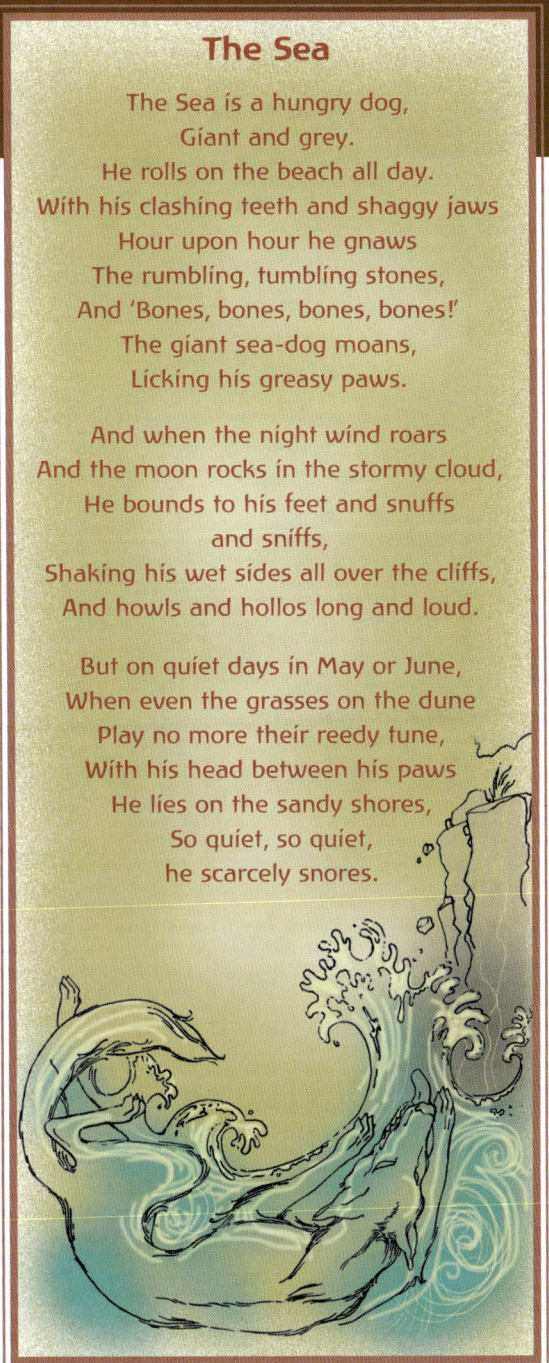

The Sea

The Sea is a hungry dog,
Giant and grey.
He rolls on the beach all day.
With his clashing teeth and shaggy jaws
Hour upon hour he gnaws
The rumbling, tumbling stones,
And 'Bones, bones, bones, bones!'
The giant sea-dog moans,
Licking his greasy paws.

And when the night wind roars
And the moon rocks in the stormy cloud,
He bounds to his feet and snuffs and sniffs,
Shaking his wet sides all over the cliffs,
And howls and hollos long and loud.

But on quiet days in May or June,
When even the grasses on the dune
Play no more their reedy tune,
With his head between his paws
He lies on the sandy shores,
So quiet, so quiet,
he scarcely snores.

△ **Figure 3.13** 'The Sea' by James Reeves, in which the poet compares the sea in summer and winter

Activities

Foundation

5 Draw four mini sketches that represent each process of erosion at the sea – which is which?

Target

6 Sketch and label the photograph in Figure 3.12 to explain how the sea erodes the coast and creates the landforms.

Extension

7 Identify lines in the poem that might be corrasion, attrition and hydraulic action. Write your own poem about how the sea erodes the coast.

What are the processes shaping the coast?

Process 3 - Transportation is moving stuff!

Remember King Canute who thought he could stop the sea coming in? He couldn't! The sea is on the move all the time and it is very difficult for human beings to stop it altogether.

△ **Figure 3.14** Abermawr, Pembrokeshire, Wales, is a large secluded bay with an impressive pebble bank above the beach, created after a tremendous storm in 1859. As the tide ebbs, sand is revealed.

At any one place the sea is affected by WOT:

W is for **waves** – the action of the wind on the surface of the sea. The stronger and longer the wind blows, the larger the waves.

O is for **ocean currents** – the movement of sea caused by warm water flowing to cold areas or vice versa.

T is for **tides** – the action of the sun and moon's gravity pulling the sea in different directions.

All these act together to transport eroded material called **sediment** from one place to another and sometimes back again! The stronger this flow of seawater, the more **energy** there is. More energy means more sediment is transported. More energy means bigger sediments can move.

Process 4 - Deposition is sorting and putting down

Have you ever noticed that a beach may have big cobbles and pebbles at the top, medium sized shingle in the middle and sand by the water's edge? Do you know why? The answer is energy! The sea is fantastic at sorting material. When it has a lot of energy it can shift the big cobbles. When it has less energy it can only move fine silts. So in big storms it will throw big cobbles to the top of the beach. On a quiet summer day it will only move the sand gently.

Have you also noticed that some beaches are either sandy or stony? Some areas of the coast are muddy, too, and some have very little beach at all. Why might that be? It is because some

places along the coast have more energy than others. Those where the sea has most energy will have very little deposition at all. Those where the sea has the least energy will have the finest sediment, a mudflat.

More **energy** means more erosion and transportation and less energy means more deposition. But is it this simple? When the sea is stormy it has more energy, and so it is more likely erosion will happen. Re-read the poem by James Reeves on page 49. Think about what makes the sea stormy. Don't turn over the page just yet. Think, why is a coast like it is?

▽ **Figure 3.15** Particle sizes of sediments (not to scale)

SEDIMENT	BOULDER	COBBLE	PEBBLE	SHINGLE	SAND	SILT	CLAY
Size of sediment	Larger than 256mm	64 - 256mm	4 - 64mm	2 - 4mm	0.06 - 2mm	0.002 - 0.06mm	less than 0.002mm
Photo →							

To become a better geographer, you need to ask:

- **Why is it like it is?**
  or
- **Why is it where it is?**

To answer these questions, **factor** analysis is useful. This means trying to work out possible causes and group them under headings.

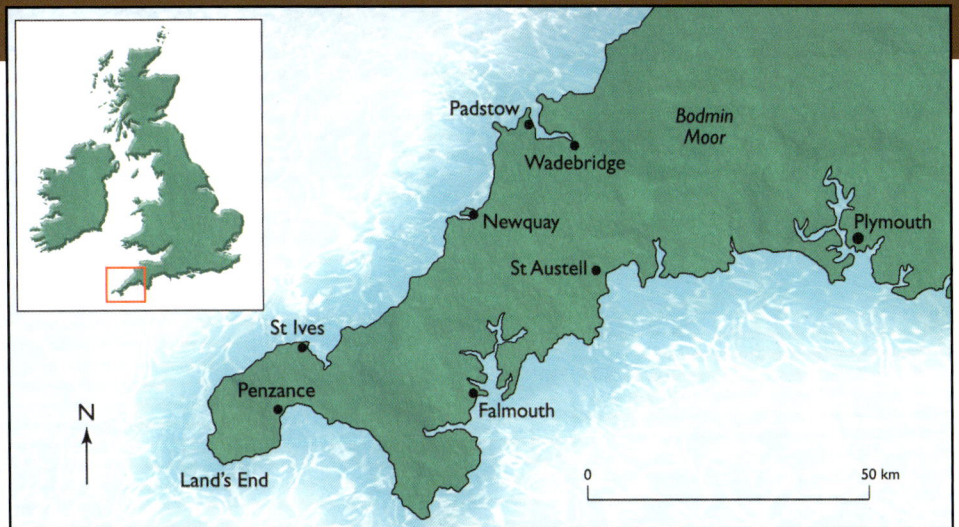

△ **Figure 3.16** Map of Cornwall with inset map of the British Isles

Activities

1 On any stretch of coast, there are a variety of factors influencing the energy available for erosion, transportation and deposition. Read the factors in the boxes opposite carefully. Then group them under the three main headings: Weather, Location and Sea Factors.

2 The map (Figure 3.16) shows four places in Cornwall. Where would you go surfing? Where would you sunbathe? Explain why? Use the boxes above to help you write down reasons.

3 Explain why Land's End has no port and Falmouth is a sheltered port.

4 Draw a sketch map to show a plan for a new yachting marina, showing the direction of the harbour wall to create a safe haven for yachts. Use the following headings – Climatic factors, Location factors and Sea factors – to explain your decision.

The windier a place is, the more likely there will be erosion.

The deeper the water, the more power to erode.

The more often the wind blows from one direction (prevailing wind), the more likely there will be erosion on exposed coasts.

The more sheltered the coast is from prevailing winds and currents, the less erosion.

The faster the speed of the current, the more erosion.

The longer the **fetch**, the bigger the waves. The fetch is the distance across the sea, to the next piece of land, in the direction of the prevailing wind.

Why is a coast like it is? – a European-wide view

The storminess of the sea is only one factor. The characteristics of the land are a factor too. The coasts of Europe are many, varied and beautiful.
Let us look at two stunning landscapes (Figures 3.17 and 3.19).

A coastal **geomorphologist** is like a toddler – he or she keeps asking 'why?' Other people who are interested in landscape also want to know why it is like it is.

△ **Figure 3.18** Map of Europe indicating coastal types

› Where can tell you why

A geomorphologist will first identify and describe the different shapes and patterns in the landscape. These are the landforms.

The stunning image of the beautiful eastern coastline of Iceland (Figure 3.17) is a special place. It is known as a **fjord**. A fjord is a deep-sea inlet surrounded by mountains on three sides where the sea goes a long way inland. There are fjords located on the west coast of Iceland and also in Norway.

› What is happening can tell you why

◁ **Figure 3.19** The Island of Zakinthos is in the Ionian Sea, part of the Mediterranean Basin. Cliffs of soft limestone surround the shipwreck

To explain the shapes and patterns of landforms a geomorphologist asks which *processes* of *weathering*, *erosion*, *transportation* and *deposition* may have been involved in the past. For example, in Iceland, valley **glaciers**, massive rivers of ice from the last ice age, eroded very deep **u-shaped valleys**. The sea level rose to flood them to create fjords.

What the environment is like can tell you why

To work out why these processes are the main ones, you need to understand the factors that make that place like it is. You can CRASH analyse it!

	Factor	Explanation	Example analysis – an Icelandic fjord
C	C is for **Climate**	How the climate (temperatures, precipitation and sunshine) affects the land near the coast.	• Cold wet climate in summer, freezing in winter. • Wet climate from the Atlantic. • Snow on mountains in winter makes beautiful coastal scenery.
R	R is for Rocks and **Relief**	The character (hardness, **permeability, porosity**, solubility in salt water) of the rock and the shape of the landscape (altitude of coast).	• Very hard old rocks with few cracks that wear away very slowly create rocky shores. • Very little sediment due to hard rocks. • Mountainous region making steep slopes and cliffs. • Deep **glaciated** valleys carved out in the last ice age.
A	A is for Air and wind	How wind speeds and direction influence wave energy.	• Wind from westerly and north-westerly direction means stormy coast but fjords are sheltered from all directions by high mountains. • Fetch is long from the north-west, west and south-west out into the North Atlantic.
S	S is for Sea movements	How tides, currents and long term sea level changes affect the character.	• Sea level risen to flood deep glaciated valleys. • North Atlantic Drift (a branch of the Gulf Stream) brings warmer water.
H	H is for Human activity	How humans are managing the coastal margin and the sea near the coast.	• Very little human impact as population is very low. • Good for sheltered ports. • Salmon farming.

△ **Figure 3.20**

Activities

Foundation

1 Annotate both photographs (Figures 3.17 and 3.19). To do this, label and describe different natural and human parts of the landscape.

Target

2 Use CRASH analysis to annotate Figure 3.19. An atlas will help. You may need to look up more details using websites or reference books.

Extension

3 Research another coastline from another continent and CRASH analyse it. You could find out about coral islands or mangrove forests in tropical zones.

Why is a coast like it is? – a European-wide view

Which landforms arise from erosion – dominated coasts?

What can be more beautiful than a coastal **arch**? All round the UK coast are features like the one in Figure 3.21. The erosion that makes cliffs steep can also create beauty. In certain places from Iceland to France to Portugal, there are features like this.

▷ **Figure 3.21** The coast of the Algarve faces into the Atlantic Ocean. These caves and arches at Praia da Rocha will turn into stacks and stumps

We spent a day on the beach in a little golden sandy cove with steep cliffs and lots of rock pools. It was a hot day and good for sunbathing but Dad can't stay still for more than a minute. He took all of us children to explore the rock pools. We had to walk through an arch and climb up onto a slimy rocky area. The pools were teeming with life. The golden sand was soft and fine but when you looked closely it was full of shells, all the colours of the rainbow. The tide came in and we jumped the waves and the arch was surrounded by water. My Dad took me on a pedalo and we explored the towering rocks protruding like icebergs. We pedalled under arches and into deep dark caves.

I had a wonderful time and I hope I can go back someday.

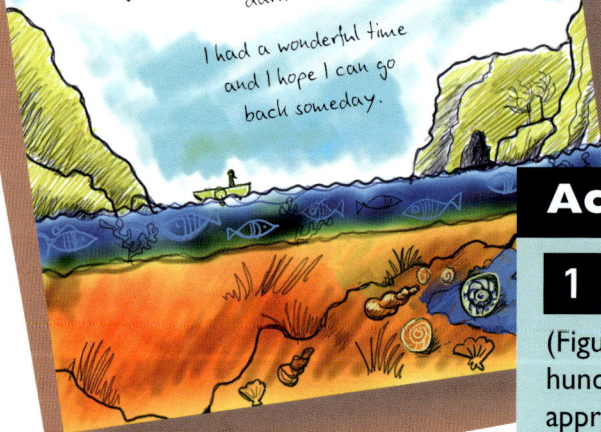

△ **Figure 3.22** My beach experiences in the Algarve, by Naomi

Ingredients
- Headland with cliffs made of medium hard rocks with lots of joints in the rock
- Deep Sea
- Wind

Instructions

Step 1   Blow the wind over the surface of the sea to make big waves.

Step 2   Allow the waves to crash into the rocks to widen cracks in the rocks.

Step 3   Allow the sea to crash into the cracks for hundreds of years so they become caves.

Step 4   Ensure a crack on the other side of the headland is attacked too.

Step 5   Keep allowing the sea to attack both caves until they meet creating an arch.

Warning! Unless erosion and weathering is slowed down, the arch will get bigger and bigger until it collapses. Then you will have a stack. Or worse still, the stack will be washed away leaving a stump.

△ **Figure 3.23** Recipe for an arch on the coast

Activities

1 Sketch the photograph in Figure 3.21 and label the features you can see. Using the recipe for an arch (Figure 3.23), sketch how Praia da Rocha will look in a few hundred years. You should include a stack and a stump. Use appropriate adjectives for shape, size, colour, texture and pattern to create a wordscape.

2 Using the description by Naomi who enjoyed her ride on a pedalo, can you make the description even more geographical? Use the correct terms for the landforms and add more detail about how they were formed.

❯ When coasts erode, what happens to all the material?

Some material falls to the seabed to form the rocks of the future. This is called **sediment**. When the rocks become hardened by pressure of more sediment on top, they become sedimentary rocks. Not all of the material falls to the deep sea. Figure 3.24 shows Spurn Head. This **spit** is one of the wondrous landscapes of the UK. But how did it come to be? Figure 3.25 below gives us a clue.

△ **Figure 3.25** Further up the coast from Spurn Point is Hornsea. Note the beach either side of the groynes.

Activities

3 You can find other spits along the coast of the UK, and Southwest France also has some. Locate them on maps of Europe. Can you find others across the world? How long are they?

- Blakeney Point, Norfolk
- Hurst Castle Spit, Hampshire
- Orford Ness, Suffolk
- Cap Ferret, Les Landes

Key terms for the diagram in Figure 3.25
Swash – washing sediment up the beach
Backwash – washing sediment back down the beach
Groynes – wooden barriers pointing out to sea

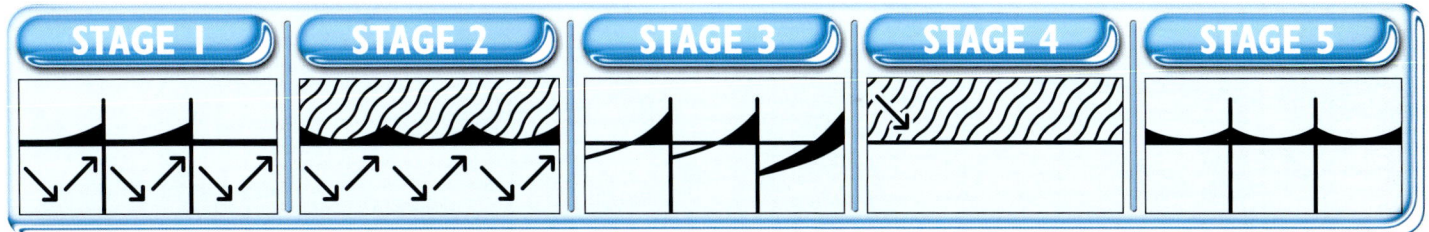

STAGE 1 | STAGE 2 | STAGE 3 | STAGE 4 | STAGE 5

△ **Figure 3.26** These diagrams illustrating longshore drift are out of sequence

Activities

4 Figure 3.26 tells the story of Longshore Drift, but it is out of sequence. Put the explanation diagrams in the correct sequence and number them. Write a caption for each of the stages.

only for each diagram. You will need to use the key terms.

5 Imagine this diagram is an animation for a website or a TV programme. Write a script for it. One or two sentences

Extension

6 Using Figure 3.23, draw a storyboard for an animation to explain how an arch is formed, using all the terms for the landforms and explain in more detail about the erosion processes. Remember CASH from page 49.

ICT activity

Scan a photograph of a coastal landscape (like Figures 3.21 or 3.24) and go forward a thousand years to show how it will change. You can do this with a computer graphics package. What might happen to Spurn Point if the sea level rose by a few metres?

What can maps tell us about erosion coasts?

This map (Figure 3.30) shows part of South Dorset. This coast is a world heritage site because of its beautiful and unusual landforms, like Lulworth Cove, Stair Hole and Durdle Door (Figure 3.11 on page 48). Geology and the sea combine to make this special place.

▷ **Figure 3.27** Location map of Dorset in the UK

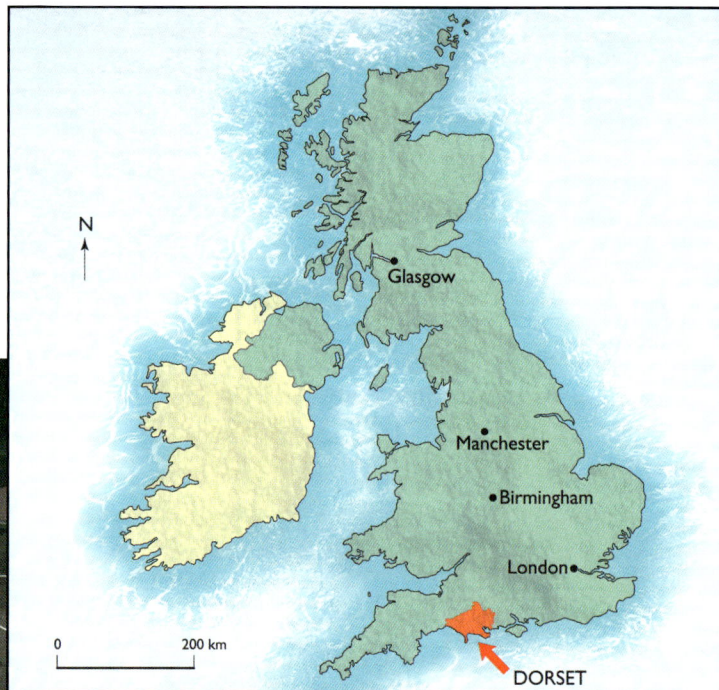

◁ **Figure 3.28** Lulworth Cove, South Dorset, a place where the sea has broken through the hard band of rock and eroded the softer rock behind, making an almost circular cove. More images can be seen at www.soton.ac.uk/~imw/dorlist.htm

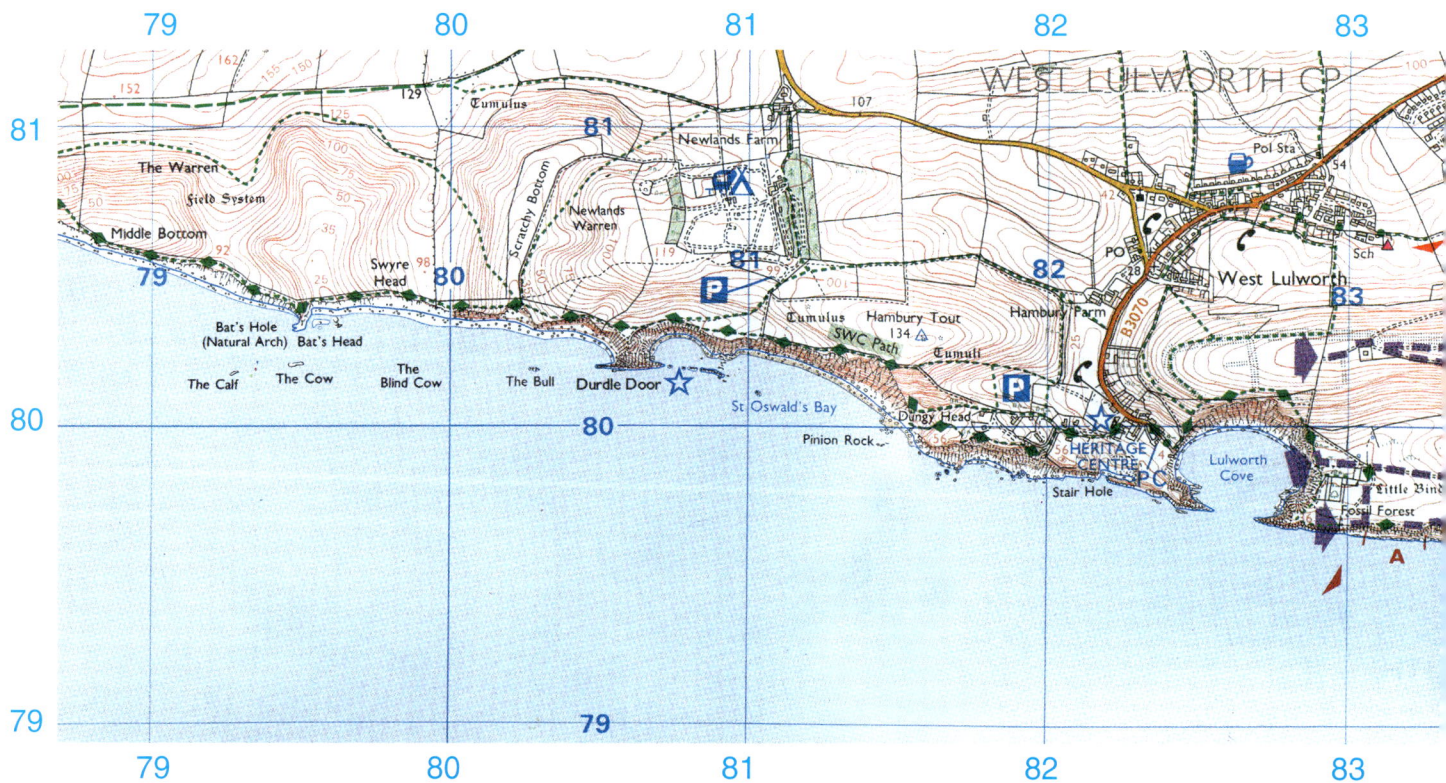

Place on map	Feature	Grid references (6 figure in brackets)
Lulworth Cove	Cove	8279 (827798)
Bat's Hole		7980 (795804)
	Cave	8379
The Bull	Stack	
The Calf	Stump	(792802)
	Arches	8279 (822798)
	Wave-cut platform	8479
Black Rock		
	Headland and cliffs	8080
St Oswald's Bay		8180 (813802)

ICT activity

Using desktop publishing and www.multimap.com, research stretches along the UK coast where people might go for a walk. Then using www.google.com find some images of that coast and put them together as a PowerPoint presentation.

Activities

1 Complete the first two columns of the table using the four-figure grid references to help you.

2 Complete the table with six-figure grid references.

3 Plan a day's walk taking in as many of these features in the table as possible. Each grid square is 1 km. Up to 12 km is quite strenuous in a hilly landscape and is probably a sensible limit.

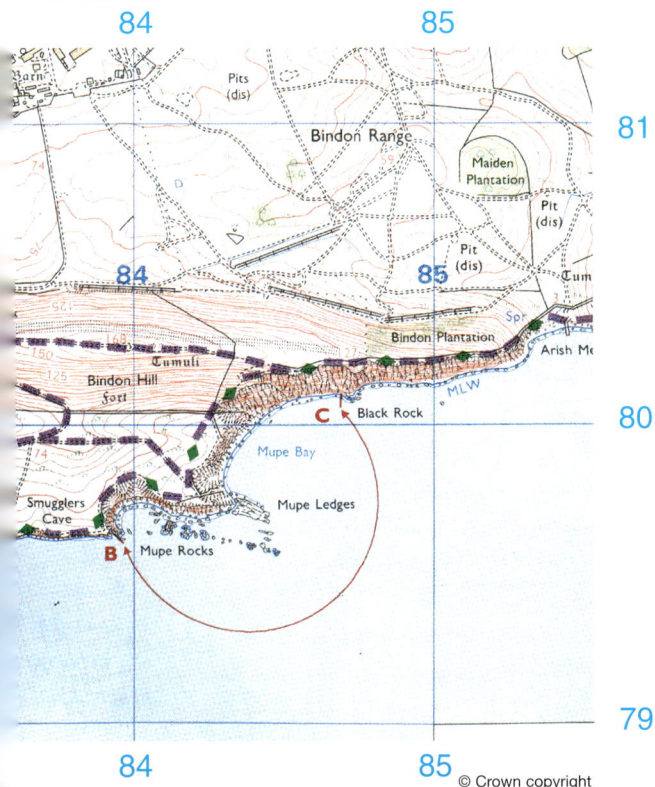

◁ **Figure 3.30** 1:25 000 extract from the Purbeck and South Dorset Ordnance Survey Explorer OL15

© Crown copyright

What patterns do maps reveal on deposition coasts?

△ Figure 3.31   ▽ Figure 3.32

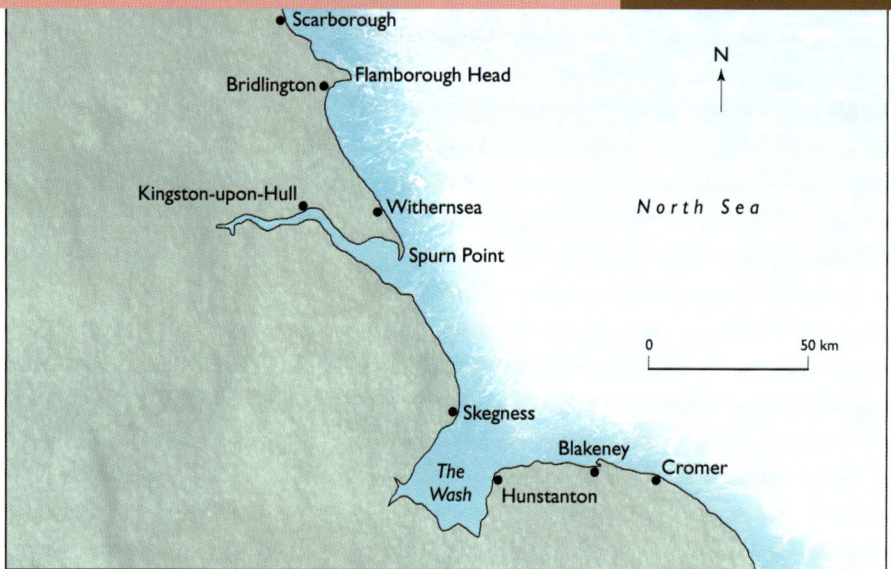

△ **Figure 3.33** The East Coast is a system of erosion, transportation and deposition of sediment. The Wash is where large amounts of it end up.

The coast has many landscapes which at first sight in photographs seem dull. On closer inspection, the map reveals patterns of habitats that are good for wildlife.

Mudflats are teeming with wildlife. Curlews (Figure 3.31) use their long beak to probe the mudflats for worms and other invertebrates.

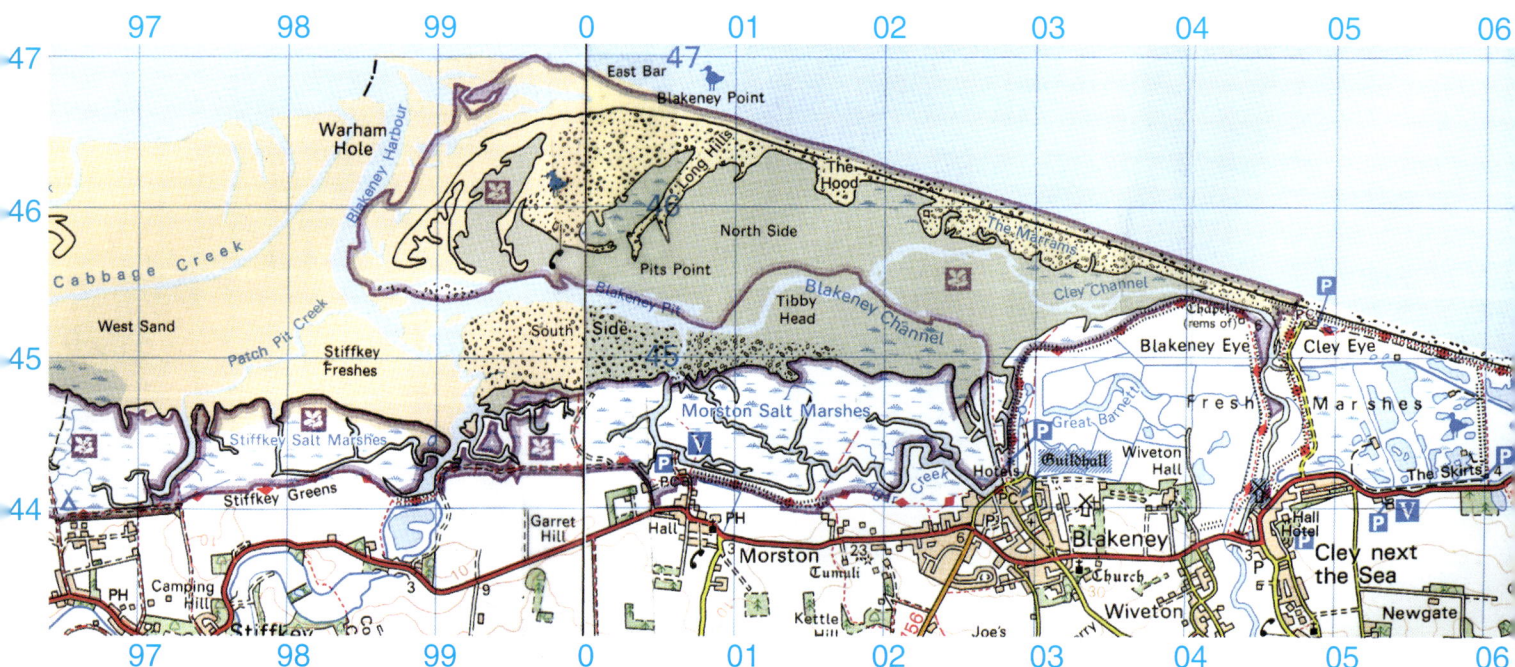

Seals (Figure 3.32) enjoy sleeping on sandbanks. They are shy of people so when it comes to giving birth they need land away from people.

Millions of wading birds winter on the sediments of the East Coast of the UK. The Ordnance Survey map below shows the North Norfolk Coast. Maps can be used to plan a trip to watch nature.

△ **Figure 3.34** Blakeney Point, North Norfolk, an extract of Ordnance Survey Landranger 1:50 000 maps

© Crown copyright

Activities

Foundation

1 Complete the table:

Place	Landscape	Grid reference
Warham Hole		985465
	Creek in mudflats	015456
Long Hills	Stony beaches	
	Freshwater marshes	055442
Stiffkey Marshes		
		030458

Target

2 The table below shows wildlife and the habitats they like. Imagine you are visiting Norfolk for a wildlife-watching weekend and so you have to decide where it is best to see the following. You may need a key from a 1:50 000 Landranger map.

Wildlife	Habitat	Grid reference
Seals	Sandbanks for breeding and resting	
Curlew	Mudflats, for feeding on invertebrates in mud	
Turnstone	Stony beaches for feeding	
Bittern	Freshwater marshes with reeds for nesting and shelter from predators	
Samphire	Saltmarsh because the plant can grow in salty soils	
Marram grass	Dunes because the grass likes to grow in free draining sandy soils	

Extension

3 The landscape in the OS map (Figure 3.34) has several currents affecting it.

a Explain the map evidence for longshore drift.

b Insert in the explanation below references to specific places and grid references on the map.

How tidal currents are responsible for shaping this landscape

The pattern of shingle, sand and mud can be explained by energy in the sea water as the tide ebbs and flows. At low tide the shingle, sandbanks, mudflats, and saltmarshes are exposed. If the sea is stormy, it has more energy, and more sediment of all sizes will be moving in the water. At high tide the sea may have enough energy to throw shingle and pebbles on to the shingle beach [_ _ _ _ _]. As the tide begins to flow, it fills the creeks first [_ _ _ _ _], eroding beds and banks and picking up more sediment. When the creeks are full it spills over onto the sandbanks [_ _ _ _ _], stirring up more sediment at first but as the currents slow, it will drop the larger sand particles. Finer particles stay in the water. As the tides rise further, the sea flows further inland on to the mudflats [_ _ _ _ _], carrying the finer particles until it reaches the high water mark [_ _ _ _ _]. At high tide currents slow, the water loses energy, dropping silt and clay particles. Very high tides will cover the salt marshes [_ _ _ _ _] too. As the tide ebbs and the water gets shallower still, the sea has even less energy and can no longer carry the finer silts and clays so it drops even more on the mud flats.

Should we try and stop erosion at the coast?

This is a briefing paper from a documentary researcher for a TV producer. These are notes.

▽ **Figure 3.35**

Script Ideas – what's the issue?

Imagine owning land knowing the ground beneath your feet is falling away! The Holderness coast is eroding at over a metre a year, in some places up to three metres per year. This is the fastest eroding coast in Europe. This man (Figure 3.35) manages the Golden Sands Holiday Park for his parents. It is situated at Witnernsea and is losing land, usually occupied by caravans, to the sea. Should we try and stop the erosion? Some people think we should.

'If we don't stop the sea, it will cost millions!'

People disagree about what should be done.

'If we keep protecting the coast, it will cost millions!'

The gas terminal built 20 years ago at Corbley was thought to be very safe but now it must move. Farmland and villages are being lost. Some small towns have spent millions on sea defences. There is a big debate about it. Should we all pay more taxes to save caravan sites on the coast? Protecting a piece of farmland may be very different from protecting a town or a big tourist attraction. The sea has energy and if you stop it eroding in one place, it will choose a weaker spot!

Visuals - Film of house collapsing

Visuals
- Interview with manager of gas terminal
- Interview with local residents,
caravan site owners, farmers

Script Ideas - So what? Where is it? Why is it happening?

Holderness is on the East Yorkshire North Sea coast of the UK. This map (Figure 3.36) shows the location of Holderness, which is a flat plain made of boulder clay lying between the chalk cliffs of Flamborough Head and the Humber Estuary at Spurn Point.

Boulder clay is soft rock formed by glaciers in the last ice age. It is what it says it is, clay with pebbles and boulders mixed in. Compared with Flamborough Head's chalk, boulder clay is very soft.

Some geomorphologists say erosion is nothing new; if you look at this old map you will see lost villages where there was once land. Most erosion happens when there are high tides combined with stormy weather. Others say erosion is getting worse because the sea level is rising.

Visuals - Images of wading birds in close up

The erosion is made worse because the sediment washed away is moving south along the coast towards the Humber estuary. It becomes sandbanks and mudflats further down the coast. Holderness's loss is millions of Arctic birds' gain. The East Coast, and the Wash in particular, is one of the most important sites for migrating birds in Europe. Much of the mud, silt and sand from Holderness ends up in the Wash. At low tide in winter, millions of birds go out onto the mudflats to feed on small invertebrates in the mud. These birds will fly back to the Arctic in summer.

Visuals - Animation of map to show lost villages

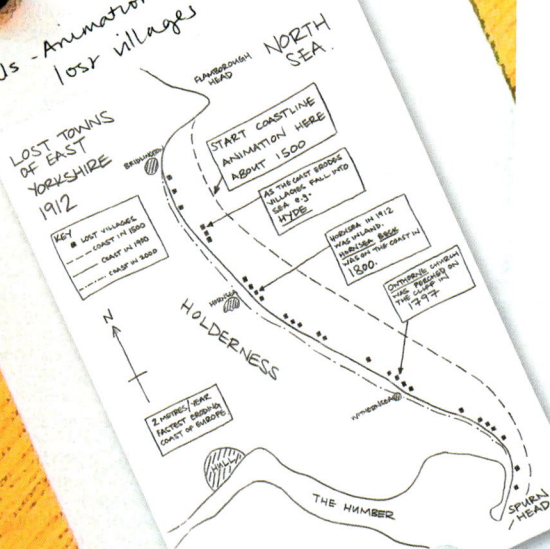

LOST TOWNS OF EAST YORKSHIRE 1912

NORTH SEA

FLAMBOROUGH HEAD

START COASTLINE ANIMATION HERE ABOUT 1500

KEY
■ LOST VILLAGE
— COAST IN 1800
— COAST IN 1900
— COAST IN 2000

HOLDERNESS

N

2 METRES/YEAR FASTEST ERODING COAST OF EUROPE

THE HUMBER

SPURN HEAD

△ **Figure 3.36**

1 **Why not role play?** You are either for protection or against it. Write a short speech explaining your reasons.

Target

2 Pick two places along the coast of Holderness and explain the differences between solutions chosen.

Extension

3 Use the information on these pages to write a short script for a TV programme on the debate about managed retreat.

Visuals - Collage of still images of the various solutions

△ **Figure 3.37**

△ **Figure 3.38**

Script ideas - what might be done?

There are several things that could be done: not all are ways people might reduce erosion. It is possible by:

1 Building curved concrete sea walls (Figure 3.37) to deflect the waves back to sea, found in towns like Bridlington, Hornsea and Withernsea. This would cost millions to maintain and is unattractive.

2 Piling large rocks along the coast (Figure 3.37). This too is expensive as rock would have to be brought a long way and in some places beaches have disappeared when this method has been used.

3 Building a breakwater (under water), which makes the sea shallower near the coast. It is impractical and very expensive.

4 Building groynes, which slow down the longshore movement of sand. A rock groyne at Mappleton has protected the village and stabilised some cliffs but according to some local people has increased erosion further south Many people think that changing one part of the coast will always impact on others. In 1991 two rock groynes and a rock revetment (retaining seawall) were built. A bigger beach was then built up between the groynes, slowing erosion. However, further south the rate of erosion has increased significantly. This is because material which is being carried south is not being replaced (it is trapped between the groynes). Therefore there is no beach to protect the cliffs. Even during a **neap tide** the sea reaches the base of the soft cliffs and erosion occurs.

5 Dumping the sand and shingle on the beach. This is the most natural solution but a bit of a lottery especially if there is a stormy winter.

6 Managing retreat of the coast by moving people (building them new houses and/or compensating them for loss of land). This is the favoured option of some planners to reduce the cost of protection. Emotionally people find it difficult to accept.

ICT activity

Use a search engine like www.google.com to find out more about solutions to coastal erosion.

Assessment tasks

How can we stay safe from the coast by understanding it better?

These activities are designed to bring together your learning through the chapter and will form the basis for your assessment.

Target task

Create a leaflet called 'A perfect day trip to ......'. The purpose of this leaflet is to inform people about the joys and the risks of the coastal landscape they are visiting. The audience is people looking for a day's walk or drive by the sea. You will need:
• An atlas
• An ordnance survey type map of a coast

• Images and information on the landscape and ecology and the activities that are possible.

It should include:
• A location map to show where it is in the UK or another country so people know how to get there
• A detailed map of the walk or car journey, with key places marked and labelled
• At least one explanatory diagram of a landform
• Warnings of dangers explained
• At least one image or drawing of the landscape.

◁ Figure 3.39

Extension task

Organise a display called 'The Coast – Beauty and the Beast' for the foyer of the school, the local library or as a special event for your school parents' evening. The purpose is to inform your local community of the delights and the dangers of different coastal landscapes. You will need:
• Maps of various coastal landscapes and atlases
• Access to the Internet for images, information and maps.

You should include:
• Maps and images of your top ten coastal landscapes in the UK and or Europe with reasons related to landscape, ecology and activities

• Diagrams to explain the dangers of coastal erosion features and coastal deposition features, perhaps including stories of people.
• If possible a presentation for an assembly using ICT.

The Coast
Beauty and the Beast

by Naomi Russell

Welcome to Beer

DANGER WARNINGS FOR LOCALS

▷ Figure 3.40

Review

What have you learnt about coasts?

1 Decide which box you feel would apply mostly to yourself.
2 Find evidence in your work and put a page reference in the box below.
3 See if your teacher agrees.

Coasts		I am confident	I have achieved	I need more help
Foundation	1 I am able to read a map of a coastal environment and identify landforms.			
	2 I am able to label a sketch, a photograph or a map with correct coastal terms.			
	3 I can describe different coastal features and identify whether erosion or deposition is the main cause.			
	4 I am able to name and locate two coastlines in the UK: one with mainly erosion landforms and the other with deposition landforms.			
	5 I am able to argue for or against the protection of the land from coastal erosion.			
Target	6 I am able to read a map of a coastal environment and interpret what the landscape looks like.			
	7 I am able to annotate a sketch, a photograph or a map with correct terms for landforms and processes.			
	8 I can describe and explain how erosion and deposition processes create different landforms at the coast.			
	9 I am able to name and locate places across Europe with similar coastlines.			
	10 I am able to describe and explain both sides of the argument for and against protecting the coast from erosion.			
Extension	11 I am able to read a map of a coastal environment and suggest explanations for the shapes and patterns on the map.			
	12 I am able to write explanation annotations for a sketch, a photograph or a map with correct coastal terms.			
	13 I can explain why there are different erosion and deposition processes along different coastlines, creating different characteristic landforms and patterns on each coast.			
	14 I am able to name and locate places across the world and make comparisons between the processes operating.			
	15 I am able to describe and explain both sides of the argument, for and against coastal erosion, and evaluate the best evidence for a particular place.			

What's the story?

GRAFFITI APPEARS AFTER POLICE SHOOT MAN WIELDING GUN

△ **Figure 4.1** Saturday 16 October, 1999, Cincinnati, USA

Activities

1 Look at Figure 4.1.
  a What is the story here?
b Who are these people?
c Who are they talking to?
d Why are they here?
e Imagine what happens next.
f Write your own story for a newspaper to go with this picture. When finished compare what you wrote with the real story.

> **Real story from the *Cincinnati Enquirer***

A life-and-death struggle for a 9 mm handgun ended in the third fatal shooting by Cincinnati police this year; one the chief of police declared 'a textbook example' of proper response. However, angry anti-police **graffiti** appeared along with tributes to the slain man, Carey Tomkins, within hours of shooting in the West End. A handful of police cruisers were in the neighbourhood Saturday afternoon, and city workers were dispatched to paint over graffiti on several store fronts.

△ **Figure 4.2** Monday 28 May, 2001, Oldham, England

Activities

2 Look at Figure 4.2.
a What action is being photographed?
b Who are the people?
c What do you think they feel about the situation?
d Write a caption to go with the photograph.

HUNDREDS OF YOUTHS HAVE CLASHED WITH POLICE DURING A WEEKEND OF RIOTING IN OLDHAM

❯ **Real story from a local newspaper**

Monday 28 May, 2001

Two nights of violence between white and Asian youths resulted in injuries to 15 police officers. 29 people were arrested. This follows prolonged periods of racial tension in Oldham. It has been suggested that it was started by members of the National Front rather than clashes between Asian youths and the police.

Activities

3 **a** Do you think both photographs show extreme situations relating to law and order?
b Do the police work like this most of the time?
c Where in towns and cities are we most likely to find situations like this?

What do you know about crime?

We watch the TV and videos. There is always plenty of crime shown. But is this just drama? Crime can be committed either against people or property. If a crime is reported to the police it is classified as a **recorded crime**. However, not all crimes are reported; these are **non-recorded crimes**. Some crime goes unreported.

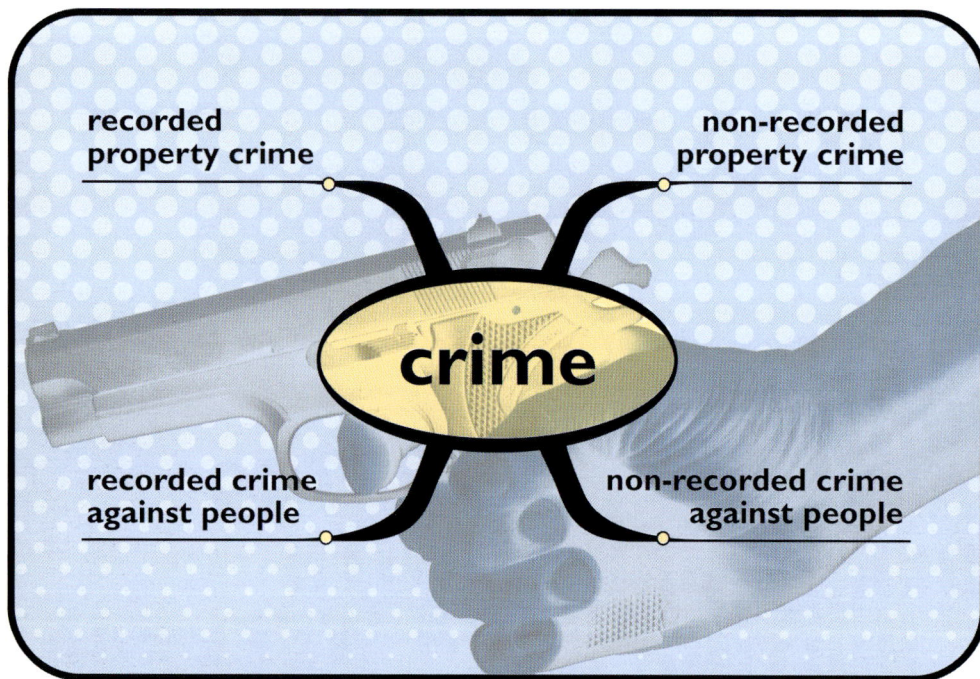

△ **Figure 4.3** Concept map – crime

▽ **Figure 4.4** Car crime: this radio was stolen from a car while the owner was taking a walk in the countryside.

Activities

1 Brainstorm a concept map, using the structure in Figure 4.3, to find out what the class knows about crime.

2 Use the concept map to suggest where crime against property and crime against people might take place.

3 Would you report the following?

- If someone did not pay their bus fare.

- If your car was broken into and the radio was stolen.

- If your bicycle pump was stolen.

> Did you know?

Even if you report a crime it may not be cleared up. 'Clearing up' means that someone is caught and punished for the offence. The crime is solved.

▷ **Figure 4.5** Percentage of crimes that get solved

	ENGLAND & WALES	SCOTLAND	NORTHERN IRELAND
Drug offences	97	100	90
Violence against the person	71	80	57
Sexual offences	68	77	76
Robbery	23	33	19
Theft and handling stolen goods	22	31	23
Vehicle theft	18	28	13
Burglary	19	21	17
Vandalism	17	23	14

Source: Social Trends 1998-9

▽ **Figure 4.6** Decorated van in California

△ **Figure 4.7** A New York subway train sprayed with graffiti, 1980s

Activities

4 Look at Figures 4.6 and 4.7. They are two images of pop art. Figure 4.6 shows a van in California USA. The owners have decorated it. Figure 4.7 is of a New York subway train in the 1980s. It is sprayed with graffiti. Are both works or art? Which one is illegal?

Foundation

5 Why do some crimes get solved more than others?

Target

6 Describe the differences between the percentage of crimes that get solved in countries of the UK. Suggest any reasons for these differences.

Extension

7 Write a crime report for your local newspaper using figures for one of the countries.

Is fear of crime justified?

ONE IN FOUR BRITONS LIVES IN FEAR OF BURGLARY

△ 23 August, 1999

Many people worry about crime in their local community. This is called **fear of crime** but usually there is much less than people think. Figure 4.8 contains some interesting statistics.

	INNER CITY	URBAN	RURAL
Vehicle theft	23.7	16.2	12.0
Vehicle vandalism	9.2	7.2	4.9
Bicycle theft	8.9	5.2	2.4
Burglary	8.5	5.9	3.4
Home vandalism	4.3	3.7	2.6
Any house offence	32.3	28.6	22.8

△ **Figure 4.8** Percentage of people who fear crime (1997 British Crime Survey)

British Crime Survey - 1997

'The area in which people live can affect whether they become a victim of crime.

People living in **inner city** areas are more likely to be victims than those living elsewhere.'

△ **Figure 4.9** From the 1998 British Crime Survey

▽ **Figure 4.10** From a government report, 'Regional Trends 1998'

18 % of females feel unsafe when walking alone compared with 3% males.

8% of people feel that their quality of life is affected by fear of crime.

Activities

1 You work for a TV company and a government minister wants clear diagrams to show during a speech called 'Crime in the Community'. Use the figures above to make some useful graphs and diagrams. Using a computer could help.

2 With this information, write a short TV script for a news report that will be read out while your diagrams are shown on the TV screen. This could be an ideal opportunity to use a computer presentation program such as PowerPoint.

Points to consider:
• What do you think the public ought to know?
• Do you think your report will make the fear of crime worse?

Some reasons for the changes

ICT activities

You have been asked to conduct a Fear of Crime report for a teenage magazine. The magazine would like to find out about teenage views, parents' views and even grandparents' views.

The magazine wants some graphs and diagrams to go with the report. You will be restricted to two pages. The magazine editor wants all the information to be submitted using IT. When completed, try e-mailing it to a friend.

1 Design a questionnaire, using a word processing package. Discuss with a partner the different types of crime you are going to ask about, e.g. mugging, burglary, rowdy behaviour. Use a 10-point scale: 0 = no fear, 10 = very great fear. Interview a mixture of males and females, and different age groups.

2 Use a spreadsheet to record the results of your questionnaire. Present these results as a series of graphs (Figure 4.13 gives an example.)

3 Write your report. You could use a desktop publishing package for this.

△ **Figure 4.11**

What do you fear?

Mugging

Theft from cars

Rowdy Behaviour

Vandalism

Assault

Burglary

Threats

Fear of Crime Survey

	Teenagers	Parents	Grandparents
Assault	7	8	9
Car theft	3	9	7

△ **Figure 4.12** Fear of Crime Survey (average scores)

FEAR OF CAR THEFT

teenagers

parents

grandparents

age groups

very high

△ **Figure 4.13** Fear of Crime bar chart

How do you find out about local crime?

One way to discover how much recorded crime is happening in your area is to read the reports in the local newspaper. You can estimate how important the paper thinks the story is by calculating the amount of print given to the story. To do this measure the area of the story in cm^2.

If you do this for a few months you will be able to see what local crime is taking place. With a local street map you could build up an interesting map as well.

Figure 4.14 is the log of local crime from a weekly free newspaper in an English Midlands town, population 90 000.

HEADLINE NUMBER	STORY	AREA CM^2	PAGE
1 **CRIME BUST BOOST**	400 council residents will have downstairs window locks fitted free as the fight against crime is stepped up	236	1
2 **Thief strikes at market stall**	A thief stole £40 cash from a stall while the assistant was busy serving a customer	23	1
3 **Blast risk as fire guts garages**	Arsonists who gutted three garages which could have sparked an explosion are being hunted by police	64	3
4 **VANDALS IGNITE RAILWAY EMBANKMENT**	Firefighters spent three hours tackling a blaze on a railway embankment after vandals ignited grass	28	3
5 **Woman's bag snatched**	A youth stole a woman's handbag from her shoulder as she walked in broad daylight	40	6
6 **PARKING BAN COULD MEAN WOMEN BEING IN DANGER OF ATTACK**	Worried residents have submitted a petition against plans to stop them parking outside their houses amid fears that they could be attacked	142	6
7 **BURGLARS DISTURBED**	Burglars stole over £1000 worth of property from a house but were disturbed when they broke into another property shortly afterwards	62	9
8 **Bike stolen**	A purple mountain bike was stolen from outside a rugby club	16	10
9 **HOAXER LASHED**	Firefighters have condemned hoax callers who put lives at risk by dialling 999	24	14

△ **Figure 4.14** Log of local crime

What is defensible space?

Some people think that crime may be related to how space is organised. Space may include your front garden or a local park. Does the space feel as if it belongs to you? Is it part of your own **territory**? Can you keep an eye on the space? We call this **surveillance**. It is thought that space that can be more easily defended has less crime. Space that looks owned and overlooked by someone is more easily defended.

△ **Figure 4.15** Photo A

▷ **Figure 4.16** Photo B

▽ **Figure 4.17** Photo C

Activities

1 What does Figure 4.14 tell you about crime in this town?

2 Look at the photographs of the two housing areas, Figures 4.15 and 4.16. Both are modern developments. Now look at Figure 4.17 of another area showing a foot tunnel under a busy road. The white wall is often sprayed with graffiti. No one lives nearby. The local council owns this area. After carefully studying all three photographs, complete the table below to summarise what you think about **defensible space** for the three photographs.

	TERRITORY	SURVEILLANCE
Photo A		
Photo B		
Photo C		

3 Think of an area you know. Write about your ideas on territory and surveillance. If you have access to a digital camera you will be able to label the main features.

Shop locally, shop safely?

Shopping is an activity that we all take part in. Often we shop locally but will we shop safely? Here are two photographs of local shops. Figure 4.18 shows a small shopping precinct serving a large estate in a large industrial northern English city. Figure 4.19 shows a village shop found 10 kilometres south of this city.

▷ **Figure 4.18** Council rented property on edge of city. Car ownership is low and teenage unemployment is high. Most local jobs are low paid. The local industries are in decline. The local shopping precinct provides a local service but prices are high and choice is limited

△ **Figure 4.19** This is a rich commuter village. Many residents work in high paid jobs in the nearby city. Car ownership is high. There are few young people living here. There are a large number of wealthy retired people. The village shop is used for convenience goods such as newspapers and bread. There is a community board for local events and ads

Activities

1 Which photograph – Figure 4.18 or 4.19 – is an **urban** scene and which one is a **rural** scene?

2 Would you feel safer shopping in one than the other? Give any evidence from the photographs to support your answer.

3 Can you give advice to local community leaders on how people could be made to feel safe in shopping areas?

4 Are there any things not shown on the photograph that you would like to find out more about?

5 To help you understand your feelings about these photographs, make a copy of the like–dislike chart, Figure 4.20. Draw a profile for each of the two shops. In each row, select the cross that is nearest to what you feel about the area. Then join the crosses up to make a profile.

6 What do the profiles tell you about these shopping environments? Compare your results with another student. Do you have similar profiles? Try to justify why you have drawn them this way.

7 Read Figure 4.21. What else do you think could be done to make it harder for shoplifters?

LIKE–DISLIKE CHART					
Quiet	X	X	X	X	Noisy
Friendly	X	X	X	X	Unfriendly
Healthy	X	X	X	X	Unhealthy
Safe	X	X	X	X	Dangerous
Tidy	X	X	X	X	Untidy
Attractive	X	X	X	X	Unattractive
Interesting	X	X	X	X	Boring
Pleasant	X	X	X	X	Unpleasant
Well-kept	X	X	X	X	Badly kept
Spacious	X	X	X	X	Cramped
Rich	X	X	X	X	Poor
Good plan	X	X	X	X	Bad plan
Varied	X	X	X	X	Monotonous
Clean	X	X	X	X	Dirty
Beautiful	X	X	X	X	Ugly
Homely	X	X	X	X	Impersonal
Colourful	X	X	X	X	Drab
Distinctive	X	X	X	X	Ordinary

◁ **Figure 4.20** Like–dislike chart

JOIN UP THE CROSSES			
X	X	X	X
X	X	X	X
X	X	X	X
X	X	X	X
X	X	X	X
X	X	X	X
X	X	X	X
X	X	X	X
X	X	X	X
X	X	X	X
X	X	X	X

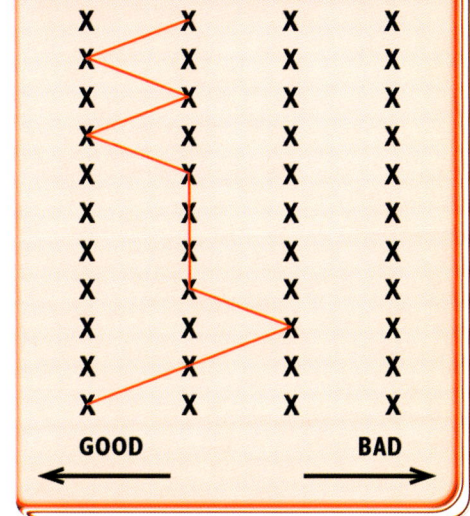

GOOD   **BAD**

◁ **Figure 4.21**

CRIME **AGAINST** SHOPS

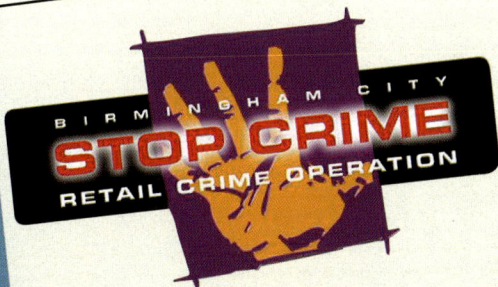

Shoplifters steal over £740 millions' worth of goods every year. Because of this we have to pay more than £10 a year extra in the shops to make up for these losses (figures from www.retailing.uk.com/report2.html)

Some shopping centres are fighting back. Birmingham has set up a group called the Birmingham City Retail Crime Operation (RCO). Some of its aims are to:

- Identify the major criminals
- Focus resources to deal with them
- Encourage the better use of information sharing
- Identify the individuals who persistently steal
- Make Birmingham City unattractive to thieves.

△ **Figure 4.22**

ICT activity

Find out more about this scheme by logging on to the website:

www.shopcrime.org.uk/pg.htm

Is there a geography of crime?

A crime needs a target and an offender. Some targets are easier than others for offenders. Targets can be located on a map so there is a geography of crime just as there is a geography of shops or industry.

Why study crime maps?

Mapping crime can help law enforcement. Police patrols can be directed to certain areas. Patterns may also help in solving crime. Information can help members of the public understand what ought to be done about a problem.

Hot spots – police need geography too!

Today many police departments use computer-mapped crime locations to show where hot spots are found. A hot spot is somewhere where there is a high concentration of crime.

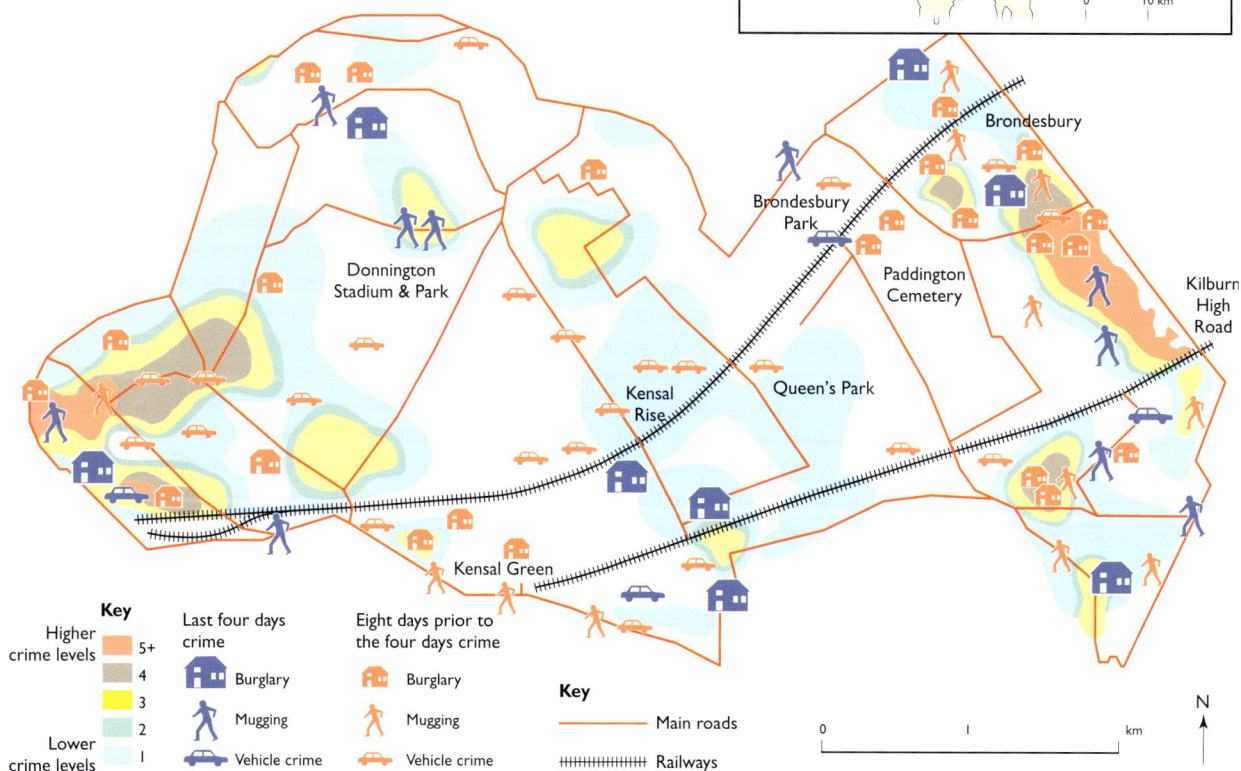

▽ **Figure 4.23** Police patrol briefing map

Key

Higher crime levels		Last four days crime	Eight days prior to the four days crime
	5+		
	4	Burglary	Burglary
	3	Mugging	Mugging
	2	Vehicle crime	Vehicle crime
Lower crime levels	1		

Key

— Main roads

++++++++ Railways

0   1   km

N

The map (Figure 4.23) shows information given to police officers in an area of London called Brent. A map is made every two weeks showing where the hotspots of crime are found. This map is issued to police officers. Areas where recent burglary, mugging and vehicle crime has taken place are shown in red or blue icons. The shading on the map shows crime trends over the last few weeks. The darker the shading, the more crime has been reported.

Activities

You are the police inspector responsible for policing the area. You have been asked by the area crime prevention panel to draw up a plan to deal with crime in this area of London. You have limited resources and you cannot have police on the beat everywhere. However, you do have four undercover detectives as well as uniformed police officers. What would you do?

△ Figure 4.24

1 Suggest a plan that would reduce the amount of crime in the area.

2 Do you think your action could move crime to other areas?

3 You will have to present your findings to a public meeting. Four different types of people will be present:

- a retired couple
- a 20-year-old who parks his car outside his flat
- a 13-year-old student who walks to and from school. She owns a mobile phone
- a 45-year-old house-owner who has been burgled in the past.

Think of a question that these locals are going to ask at the meeting and label it onto a copy of Figure 4.24. What will your answer be?

Where are the hot spots in the USA?

Does the President of the United States have to worry about crime? Washington DC is the capital of the USA. It has a very high crime rate, even close to the Capitol where the American government meets. The White House is where the American President lives and this too is close to high criminal activity.

People power

New technology is helping ordinary people report and find out about crime in their local area. In Washington you can fax or e-mail sightings of suspicious people. Should local people get involved in mapping crime? Why are they doing it? Would you get involved? Are there any problems in reporting crime in this way?

▽ **Figure 4.25** Residential burglary hot spots in central Washington DC

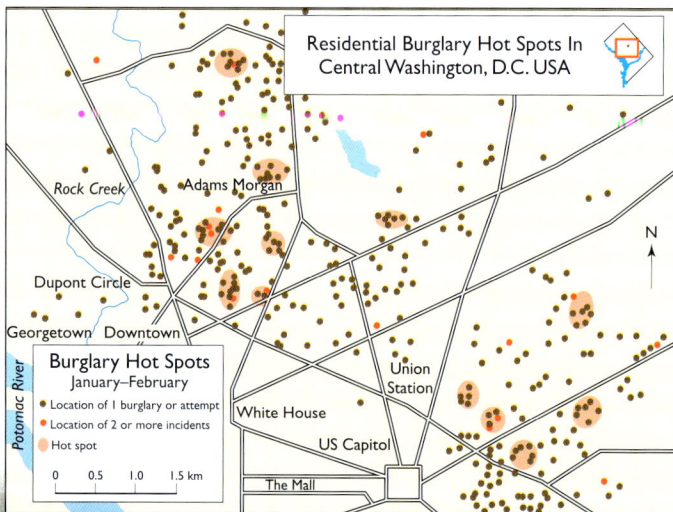

▽ **Figure 4.26** The White House

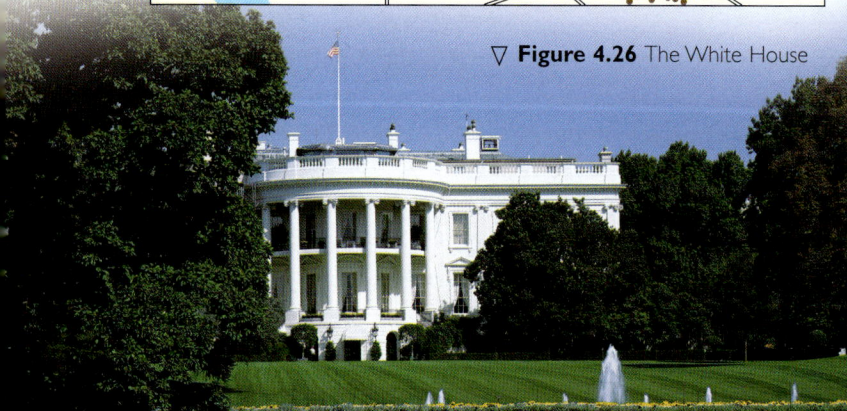

Activities

1 Draw your own hot spot map for the Washington police. You can do this by drawing a **choropleth map**. This is a map with different shadings, like the London police crime map (Figure 4.23 on page 74).

Procedure:
- Make a tracing to cover the burglary map of Washington, DC Figure 4.25. Using a ruler draw a grid of centimetre squares.
- You should have a rectangle 8 x 6 centimetres (1cm = 1km).
- Place the tracing over the map.
- For each square count the number of burglaries.
- Write the number in the square.
- Draw a neat grid (this can be done on a computer using shapes, and fill in with colour) and decide on a shading key. Remember to have the darkest shading for the highest numbers (see Figure 4.27). You will need to decide carefully on the intervals for the shading.
- Make sure you have a title and key for the map.

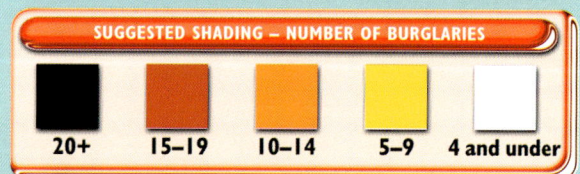

△ **Figure 4.27** Suggested shading

2 **a** Compare this type of map with a 'dot' map, e.g. Figure 4.25.

b Make a table to show the good points and bad points of using each kind of map for this information.

TYPE OF MAP	GOOD POINTS	BAD POINTS
Dot		
Choropleth		

c Which one would you use if you had to present your findings to a public meeting? Give reasons for your choice and why you rejected the other.

Putting your head in the sand

In response to America's growing crime rate many rich Americans are choosing to live in 'gated communities'. Figure 4.28 shows a gated community in California, USA. It covers 640 hectares. Facilities on offer include a golf course, clubhouse and sports facilities. It is set in southern California which has a hot dry climate. You will need to be wealthy to buy a home there.

The publicity tells you that 'Palm trees and perpetual sunshine create this gorgeous oasis in the golf capital of the world. Exceptional amenities and creative venues provide amazing variety for an active, healthy lifestyle.' To enter this community, you have to pass through a gatehouse which is 24-hour guarded. A gate is raised to let residents and visitors in. Visitors have to be checked and there are restrictions on their visit. You can find out more about it at: www.privatecommunities.com /California/SunCityPalmDesert/.

▽ **Figure 4.28** Sun City – a gated community in California, USA

Activities

1 **a** What type of people will choose to live in a gated community?

b Why would they choose to live there?

c What would you like and dislike about life in Sun City?

d Do you think the rising crime rate in the USA will encourage the building of more gated communities?

e If you were setting a gated community in Britain, what would you include and how would you organise the security?

Foundation

2 In what ways can crime be put on a map?

3 What facts do you need to make a crime map?

Target

4 Why is it often difficult to make crime maps?

5 Explain why some crimes are impossible to map.

Extension

6 Do you think the mapping of crime will help the police?

7 If you had access to a crime map would this affect where you might choose to live?

Will understanding geography help in reducing crime?

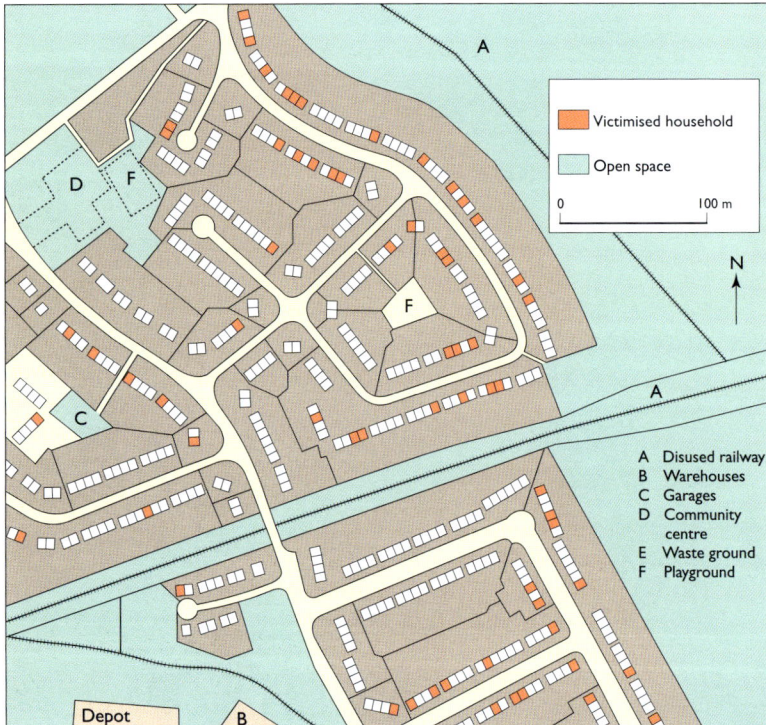

▽ **Figure 4.30** This sign shows that residents in the area are part of a Neighbourhood Watch scheme to protect it against crime

▽ **Figure 4.29** Residential area in a Midlands town showing burglaries over a three-year period

Legend:
- Victimised household
- Open space

0 — 100 m

N

A Disused railway
B Warehouses
C Garages
D Community centre
E Waste ground
F Playground

Activities

1 Figure 4.29 is a map of a housing area in the Midlands of England. It shows the houses that were burgled over a three-year period. Imagine that you live on this estate.

a Set up 'Neighbourhood Watch' groups of five or six students.

b Decide where on the housing estate you live.

c Work out which parts of the estate are worst for **burglaries**.

d Discuss what could be done to reduce the number of burglaries.

e Make notes about your top three ideas.

Activities

2 Look at Figures 4.31 and 4.32. You have been asked to advise on security. House A is on a main road; house B is in a quiet no-through road on the edge of a town. The owners want a report on what are the strong points and weak points of security. Summarise your ideas in a copy of the table below.

△ **Figure 4.31** House A

▷ **Figure 4.32** House B

	Strong security points	Weak security points	Suggestions for improvement
House A			
House B			

Reducing crime in other areas – town centres

Town centres have special problems when it comes to crime prevention. Fig 4.33 shows an area of a town centre outside of shop hours. Many people visit these areas during both day and night time. During the day shops can be targeted by shoplifters. At night people are more at risk from attack. There are few residents living here to provide surveillance.

△ **Figure 4.33**

Activities

1 What measures in Figures 4.33 and 4.34 are being used to help crime reduction?
Hint: is it really a street lamp above the shop?

2 Why are more **CCTV** cameras being fitted in our town centres?

3 What do you think are their advantages and disadvantages?

◁ **Figure 4.34**

Crime: a world view

△ **Figure 4.35** Stafford jail

Crime is common in most societies around the world but some countries have higher rates than others. One way of comparing other countries is by studying the number of people in jail per 100 000 of the population (Figure 4.36).

▽ **Figure 4.36** Number of people jailed per 100 000 of the population for selected countries

Russia	558	Australia	139	Spain	90
United States	519	New Zealand	135	France	84
South Africa	368	Canada	116	Germany	80
Hong Kong	179	Mexico	97	Japan	36
Poland	160	England/Wales	91	India	23

Activities

1 Look at Figure 4.36. Match these statements against a country on the list:
- This poor country, like many other countries, does not imprison as many people because of the high cost.
- Has the highest **incarceration** rate in the European Union.
- One of the richest countries in the world (2nd largest economy), has one of the lowest incarceration rates.
- This is the richest country in the world but has an incarceration rate five to eight times higher than most other developed countries.
- This country is now the world leader in incarceration. The rise of **organised crime** and an uncertain future has caused a **crime wave** and increased the use of jails.

Foundation

2 Draw a bar chart to show the information in Figure 4.36. Describe what your graph shows.

Target

3 The population of the USA is 280 million. Work out the total prison population. Show your workings.

Extension

4 Does having a high prison population mean that it is a more safe or a less safe country to live in? Explain your answer.

Assessment tasks

Figure 4.37 shows a telephone box on a housing estate. It is always out of order. This is not a big crime issue, such as armed robbery or drug trafficking, but for local people it is a problem. You will need to study the facts and prepare a report on how to help local people.

△ **Figure 4.37** Site X   △ **Figure 4.38** Site Y

The problem

The telephone is located on the corner of a road at site X (Figures 4.37 and 4.39). Local people without telephones find it useful. However over the years British Telecom, the company that services the telephone, is concerned that the cost of repairing damage makes it unprofitable. They want to move it to site Y (Figures 4.38 and 4.39).

Solutions

Leave it in the original position

  A British Telecom should keep fixing it.
  B **Target harden** the phone and booth (by putting in unbreakable glass and strong phones).
  C Give power to the locals to clean and report damage.
  D Install Close Circuit TV on a nearby lamppost.

Re-site the phone

  E Locate it at site Y which is on a main road away from the housing estate.

Remove the phone altogether

  F An increasing number of people own mobile phones so get rid of it.

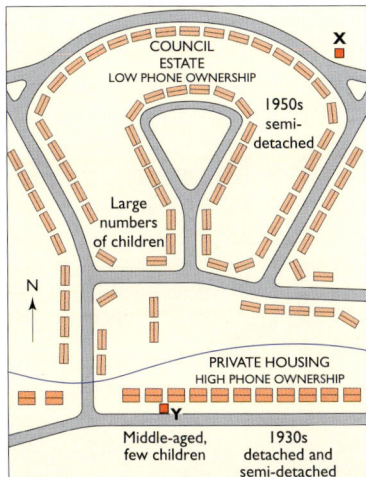

◁ **Figure 4.39** Site map

■ **X** Old site of phone box
■ **Y** New site of phone box

0   100 m

Target task

What would you do? Write in your own words what you think about each solution. Here are some ways in which you could start off your writing:

I think solution A is good because …
I think solution B is not so good because …

You must also say which one you think is the best. You could start your writing with:

I like … the best because …

Extension task

Have your say! Which is the best choice, site X or Y? Find out by giving a score for what people will think about six choices A, B, C, D, E and F. Give a score from 10 (the best) to 1 (the worst), or in between. Write down why you gave it this score. Fill in the table (Figure 4.40) and add up each choice.

▽ **Figure 4.40**

	A	B	C	D	E	F
BT manager						
Mother of two young children living near site X						
Teenager living on estate						
Middle-aged house owner near site Y						
TOTAL SCORE						

Assessment tasks

ICT activities

The Internet can give you an enormous range of information. Often there is too much! Here are a few websites that might help you to understand this topic further:

www.statistics.gov.uk/statbase/ss.asp plenty of data on British crime.

www.bromley.gov.uk/crime/index.htm a crime survey, Bromley, London.

http://courier.evansville.net/crime/weeklycrime.html a weekly crime map of a US city.

www.washingtonpost.com/wp-dyn/metro/crime Washington Post site can select an area of Washington DC to find out about recent criminal activity.

www.securedbydesign.com/ a British site that gives advice on security.

www.upmystreet.com/ find out about where you live, including crime.

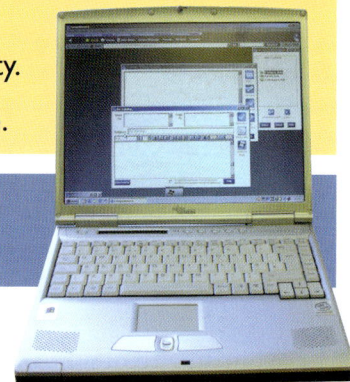

Crime without frontiers: the rise of the World Wide Web and crime

Criminal activity now extends beyond the boundaries of countries. Such activities can include drug smuggling and fraud. Increasingly computers are involved. Here are three ways in which computers are being used by criminals:

△ Figure 4.41

1 Information can be stolen or hard drives can be damaged by **computer viruses**, with the result that Internet service providers can be forced to shut down, and millions of dollars can be lost around the world.
2 The computer can be used to commit a crime such as using the Internet to sell drugs or downloading illegal images.
3 Computers can store illegal information.

△ Figure 4.42

▽ Figure 4.43

Activities

1 How do computers encourage the growth of international crime?

2 What measures should be taken to combat computer crime?

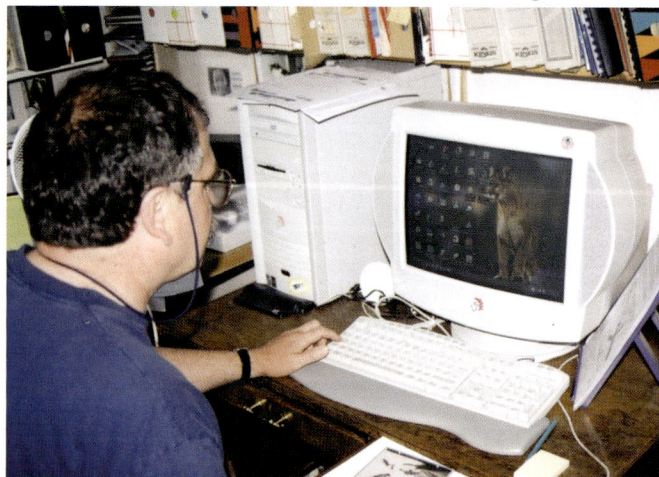

Review

So you think you know about geography and crime! Test yourself on these ten questions.

Activities

1 Do you know the difference between reported and non-reported crime?

2 Crime can be against the person or against property. Put these crimes in the correct box:

- burglary
- **mugging**
- **assault**
- shoplifting
- vandalism
- rowdy behaviour
- threats
- car crime.

CRIMES AGAINST PERSONS	CRIMES AGAINST PROPERTY

3 Why is some crime less reported to the police?

4 Which crimes are less likely to be solved? Do you know why?

5 Do we worry too much about crime (fear of crime)?

6 Do certain ages of people fear crime more than others?

7 Does where you live affect your fear?

8 Read this list of terms:
- **defensible space**
- **territory**
- **surveillance**
- **CCTV**
- **gated communities**

Either write a sentence to explain in your own words what you think they mean or write a short piece of writing using these terms in a Crime Prevention pamphlet. You may want to include any sketches, maps or photographs.

9 What do you know about crime worldwide? Which countries have more people and which have less sent to jail than Britain? Why are there differences?

10 Will computers increase crime rates or just change the type of crime? How do you think computers can be used to make it easier for the police to solve more crimes?

How well did you do? If you still want to find out more about geography and crime, be alert. Read the papers and search the Internet for more information on this important subject.

Who, or what, is Proteus?

▽ **Figure 5.1** Proteus – 'Mysterious ruler of the deep'

FACT FILE:
PROTEUS

Originally thought to be a dragon	Colourless due to lack of sun radiation
'Human fish'	**Eyes covered with a layer of skin**
Sexually mature at the age of 14	Body sensitive to light
Can survive for 12 years without food	**Amphibian**
Largest cave animal in the world	Breathes through gills and skin
A vertebrate	**Lives underwater**
Approximately 30 cm in length	Only found in a limestone region stretching from Slovenia to Herzegovinia
Lives to be 100 years old	
Lives in temperatures 8–10° C	

Activities

1 Imagine trying to survive in a world of complete darkness, high **humidity** and cool temperature.

a Why would you need **gills** and lungs?

b How would you find your food if you cannot see it?

c What shares your **habitat** and could be a food source?

d Could you find your way around? Why would you need to move around?

Scientists are still unsure how Proteus **evolved**, where it came from and why it ended up in the caves. We are still unclear on how Proteus reproduces but it does lay eggs, which hatch out into tadpoles.

What are rocks?

Rocks are made up of different **minerals**. Most minerals are a mixture of different elements. Almost 99% of the earth's crust is made up of just eight elements. These are: oxygen, silicon, aluminium, iron, calcium, sodium, potassium and magnesium.

Geographers **classify** rocks according to how they were formed. There are three main groups of rocks: **igneous**, **sedimentary** and **metamorphic**.

Igneous rocks form from hot, molten rock.

◁ Figure 5.3

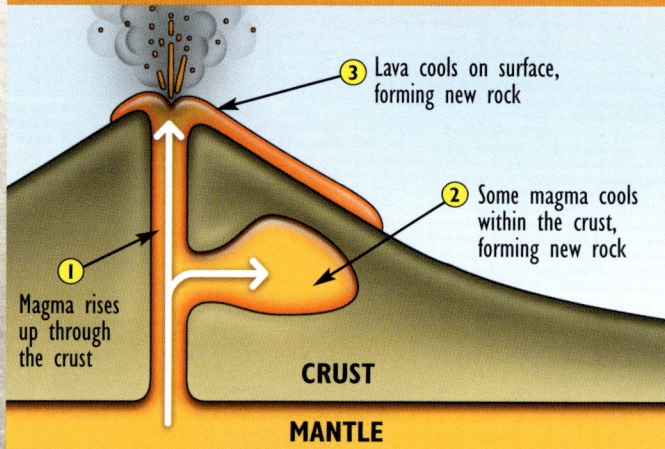

③ Lava cools on surface, forming new rock

② Some magma cools within the crust, forming new rock

① Magma rises up through the crust

CRUST

MANTLE

▷ Figure 5.2

Sedimentary rocks form from deposits of other rocks and can contain fossils of plants and animals.

▷ Figure 5.4

① Rivers wash sediment down to the sea

③ The lower layers are compressed (squashed) to form rocks

② Layers of sediment build up on seabed, along with plant and animal remains

△ Figure 5.5

Metamorphic rocks form from other rocks which have been changed by heat and pressure.

▷ Figure 5.6

◁ Figure 5.7

MOUNTAINS

PRESSURE   PRESSURE

↓   ↓

Rocks are buried and changed by heat from below and pressure from above

CRUST

HEAT   MANTLE   HEAT

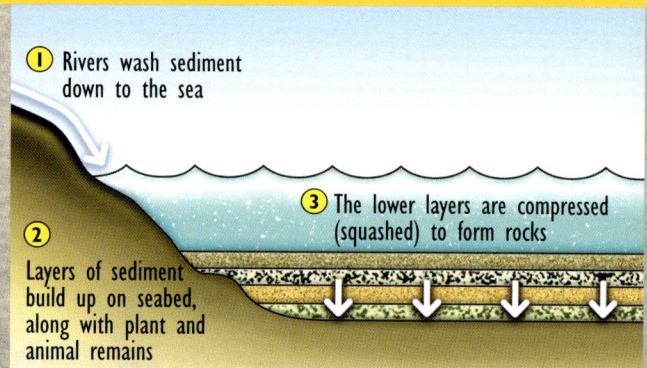

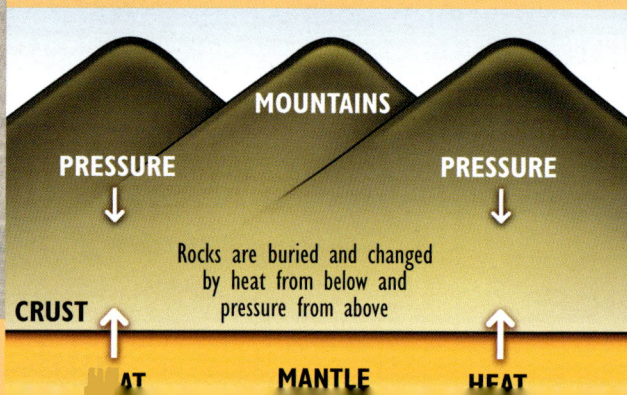

How do rocks influence landscape?

The landscape we see around us today depends upon the type of rock lying beneath.

The rocks under our feet can influence:
1 Height above sea level
2 Soil type and depth
3 **Slope** of the land
4 Vegetation
5 Water and drainage

▷ **Figure 5.8** **Granite** landscape, Bodmin Moor, Cornwall

△ **Figure 5.9** **Chalk** landscape, Etretat, France

△ **Figure 5.10** **Sandstone** landscape, Bryce Canyon National Park, Utah, USA

▷ **Figure 5.11** **Limestone** landscape, Malham Cove, Yorkshire

Activities

1 Choose one of the photographs and describe the landscape in 50 words. Compare your description with another member of your class. Can you recognise which landscape they have described?

What is limestone?

In this chapter we focus on sedimentary rocks, and in particular, limestone.

Limestone is made from **calcite**, a mineral that dissolves easily. It forms in shallow, tropical water. The calcium carbonate in limestone is from the remains of **sea shells** and **corals**.

Limestone in Britain

Around 300 million years ago, during the **Carboniferous period**, parts of Britain were covered with a shallow **tropical** sea, ideal conditions for **corals** and **crinoids**. Some of the limestone formed as reefs, from growing coral which secreted calcium carbonate, like you find on tropical islands today. **Reefs** are very hard, as they are made from dead coral cemented together, and so are not easily eroded by water and wind.

Carboniferous limestone is one of the hardest types of limestone and is grey in colour. It is mainly found in the North of England, in areas like the Yorkshire Dales **National Park** and in the White Peak area of the Peak National Park.

Britain was mainly covered in desert during the **Permian period**, except for the Permian Sea. **Evaporite limestone** formed in this sea, e.g. **oolitic** limestone.

Jurassic limestone is found between Hull on the Humber estuary and Bristol on the Severn estuary. It can also be found between the Severn estuary and Lyme Bay on the south coast of England. Jurassic limestone is softer than Carboniferous and more yellow in colour and was formed at a time when most of Britain was covered in shallow warm seas ideal for **ammonites** and corals to live.

During the **Cretaceous period**, the North West of Britain was uplifted to form mountains. The South East was covered in a shallow sea which was the ideal environment for tiny **coccoliths** to live in. These creatures helped to form the chalk beds of today's South Downs in Sussex.

▽ **Figure 5.12** Distribution of limestone in England and Wales

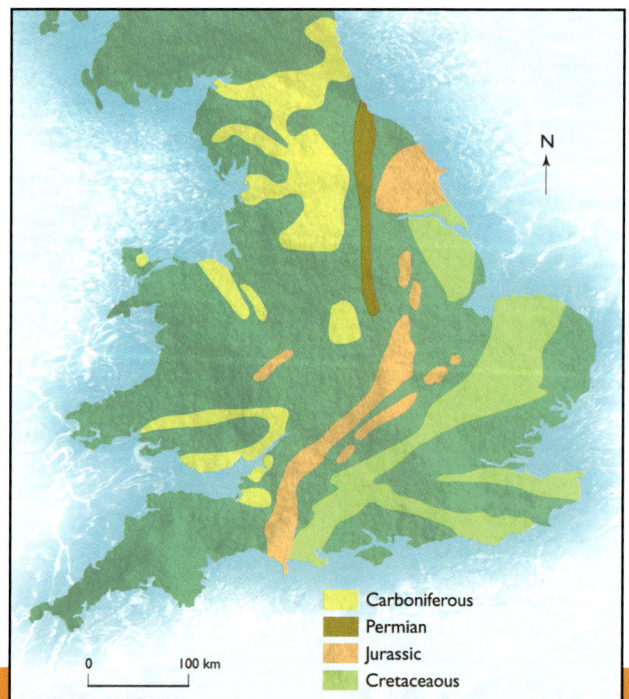

Carboniferous
Permian
Jurassic
Cretaceaous

0   100 km

Activities

1 Find out about the geological timescale. On it place the following ages of limestone in order of age, the oldest first: *Jurassic, Carboniferous, Cretaceous and Permian.*

2 Describe the location of the Permian limestones.

3 Which age of limestone is the nearest to your school?

How does limestone form?

▽ **Figure 5.13** Creation of sedimentary limestone. These diagrams are out of sequence.

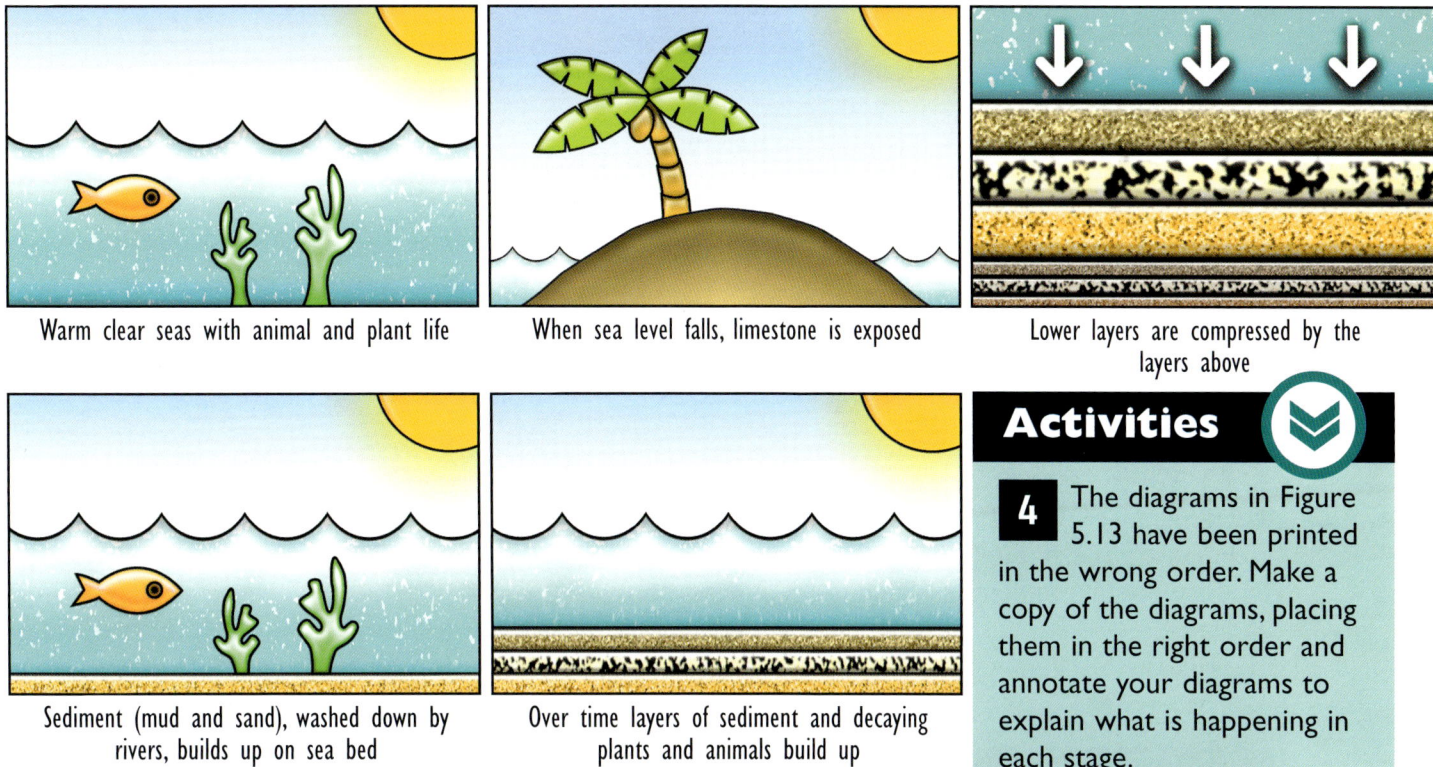

Warm clear seas with animal and plant life

When sea level falls, limestone is exposed

Lower layers are compressed by the layers above

Sediment (mud and sand), washed down by rivers, builds up on sea bed

Over time layers of sediment and decaying plants and animals build up

Activities

4 The diagrams in Figure 5.13 have been printed in the wrong order. Make a copy of the diagrams, placing them in the right order and annotate your diagrams to explain what is happening in each stage.

How to identify rock types

We can use some simple tests to help identify rock types. We can examine the following features:

- Particle size
- Hardness
- Streak
- Calcium carbonate content
- Presence of fossils

RESOURCES NEEDED:

Samples of rock e.g. limestone and sandstone

Copper coins (2p)

Tile

Hand lens

Dilute hydrochloric acid

Safety goggles

Paper to protect desks

(much of this equipment can be borrowed from the Science Department)

〉 Particle size

Study the individual grains that make up the rock. Are they:

Over 5 mm in diameter? = Coarse
Between 1 mm and 5 mm? = Medium
Smaller than 1 mm? = Fine

〉 Hardness

Different minerals have a different hardness. Geologists use a scale from 1 to 10, called Moho's scale.

Talc   **to**   **Diamond**

1 2 3 4 5 6 7 8 9 10

A mineral of hardness 7 will leave a scratch on a mineral of 6, etc. We can use the following indicators of hardness:

2.5 can scratch with a fingernail
4   can scratch with a copper coin
5   can scratch with glass
5.5 can scratch with a knife
6.5 can scratch with a steel file

How does limestone form?

› Streak test

Using the back of a tile, scratch the rock across the tile. A line of rock powder will be left. The colour of the line indicates the minerals present, e.g. Hematite leaves a red streak.

› Calcium carbonate test

Calcium carbonate fizzes in a reaction with dilute hydrochloric acid. Placing one drop on the rock surface will help you to decide if the sample contains calcium carbonate.

› Fossils

The remains of past life-forms are only found in sedimentary rocks. Studying the fossils indicates the type of environment in which the rock was created.

Activities

Identifying limestone from a selection of rocks.

Particle size = <2mm
Hardness = 3
Streak = White
Calcium carbonate = Fizzes
Fossils = Yes

◁ Example of a student's results

Foundation

1 **a** Carry out the five tests on two samples of rock. Find out which is the limestone.

b Copy and complete the table:

	ROCK A	ROCK B
Particle size		
Hardness		
Streak		
Calcium carbonate		
Fossils		

c Write a paragraph describing how you decided which was limestone.

Target

2 **a** Identify limestone from five samples of rock.

b Write up the method.

c Create a table of results.

d Justify your choice – which is limestone?

Extension

3 **a** Identify limestone from ten rock samples.

b Write up the method.

c Create a table of results.

d Discover the limitations of the techniques.

e Justify your choice for limestone.

What are limestone landscapes like?

We know that the underlying rock type influences the landscape. Different types of limestone produce different landscapes. Here are two examples:

▽ **Figure 5.14** Chalk cliffs – Seven Sisters

▽ **Figure 5.15** Limestone pavement at Malham Cove

Chalk

Chalk is the softest type of limestone; there is some chalk on the north-east coast, near Hull. Most of the chalk is found between East Anglia and the south coast, near Weymouth, see Figure 5.12 on page 88. Chalk creates rolling Downs with grass vegetation. Where it meets the sea it creates vertical cliffs (see Figure 5.14).

Chalk is made from coccoliths, the skeletons of tiny (0.003 mm) spherical marine algae (plants). One litre of sea water can contain 1 million coccosphere. The skeletons collect on the sea bed making a bed of chalk.

The tiny grains of spherical skeletons leave air gaps between them. This makes the rock porous, allowing water to soak into it through the air gaps. This characteristic of chalk explains why there are very few streams or rivers on the surface of a chalk landscape as water soaks underground.

During the Ice Age when the ground was frozen, flowing water did shape river valleys into the chalk. Today these are dry valleys, such as Devil's Dyke in Sussex.

Karst

The features of a limestone landscape are sometimes referred to as '**karst**'. This is a German word. The original word was 'Kras' which is a region in Slovenia, where many scientists carried out studies. Karst usually refers to landscapes in areas of Carboniferous limestone. Carboniferous limestone was formed around 300 million years ago.

ICT links

www.yorkshire-dales.com/welcome.html

www.mineralstech.com/limestone.html

Institute of Karst Research
www.zrc-sazu.si/izrk/

The features of limestone **scenery** have different names around the world, e.g. a **swallow hole** is called a cockpit in Jamaica and a doline in Yugoslavia.

▷ **Figure 5.16** Devil's Dyke

What is karst scenery?

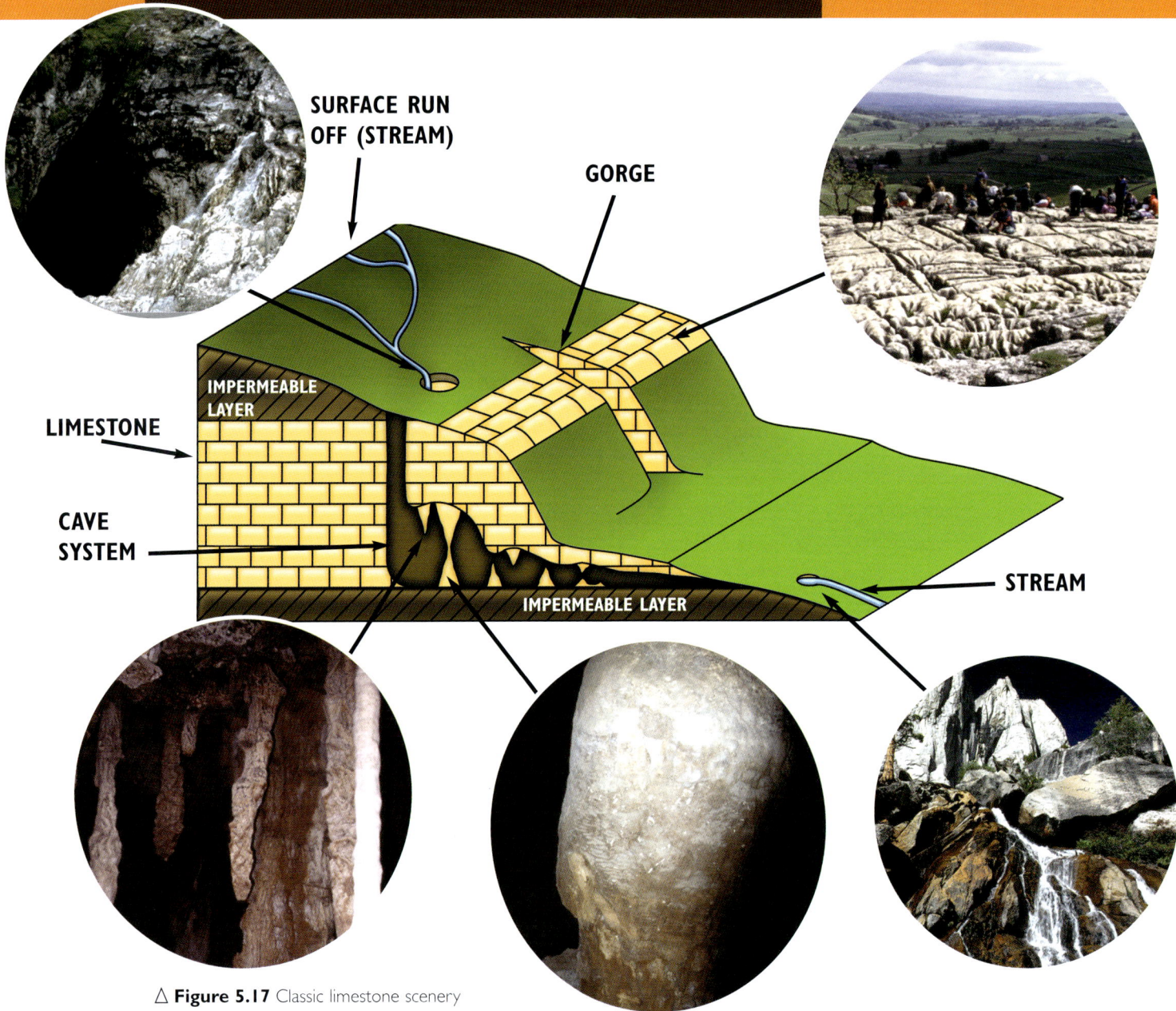

SURFACE RUN OFF (STREAM)

GORGE

IMPERMEABLE LAYER

LIMESTONE

CAVE SYSTEM

IMPERMEABLE LAYER

STREAM

△ **Figure 5.17** Classic limestone scenery

Activities

1 In pairs, study Figure 5.17 and generate three questions that you would like answering; for example, why do springs appear?

2 Copy out the table on the right. Complete the table by matching each feature shown in the diagram with a key term. Add your own description/drawing to describe the feature.

KEY TERM	DESCRIPTION
1 pavement	
2 swallow hole	
3 stalactite	
4 pillar	

3 In pairs, discuss why the following features have been given their names:
- swallow hole
- pavement.

How did karst scenery get like this?

Limestone produces a special landscape because the rock has certain characteristics.

Limestone has **bedding planes** and **joints** that are at right angles to each other (Figure 5.18). Limestone is a **pervious** rock, which lets water in along the joints and bedding planes.

△ **Figure 5.18** The joints and bedding planes in limestone

▽ **Figure 5.19** Acid rain

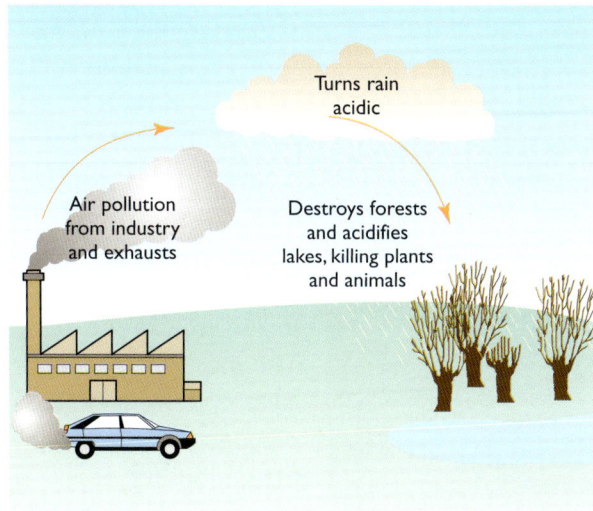

▽ **Figure 5.20** The Winnats Pass within the White Peak area of the Peak National Park

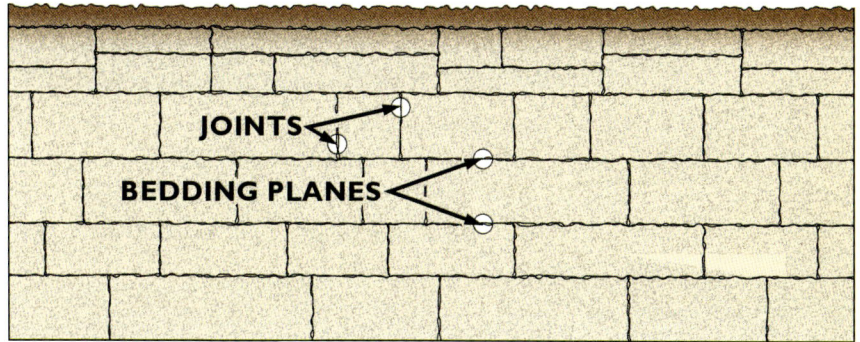

The water can find a route underground through the weaknesses, creating **caverns** and leaving **dry valleys** behind. Sometimes, the **caves** have collapsed, leaving a deep, narrow **gorge**, as at Winnats Pass, near Castleton in Derbyshire.

Chemical weathering can also attack the surface of limestone. Rainwater can pick up pollution from exhaust fumes and factories that turn it into a weak acid. This is called **carbonic acid** and it can dissolve rocks like limestone, which contain calcium carbonate.

Activities

4 Draw diagrams for each of the following statements, to show how water can attack and dissolve limestone:

> Acid water (carbonic acid) seeps into the joints and bedding planes

> Acid rain falls on limestone

> Caves are produced underground

> The carbonic acid dissolves the rock

Write the statements beneath the diagrams to explain what is happening.

Erosion by ice has also shaped the landscape. Between 100 000 and 10 000 years ago parts of Britain were covered with huge rivers of ice, called glaciers. At the end of the Ice Ages, temperatures rose creating melt water. The melt water, carrying rock debris, eroded the landscape, creating gorges and caves.

Case Study – Belize, an example of karst scenery

Limestone in Central America

There are thousands of limestone caves in Belize. Belize is a country in South America. Most of the caves are found in the south and west of the country. The limestone in Belize started to form during the Cretaceous period.

The Ancient Maya people lived in this part of South America around 1 400 years ago. They are famous for being the first people to make use of chocolate. The actual word 'chocolate' is derived from the Mayan word *xocolatl*, which translates as 'bitter water'. They made a drink (*xocolatl*), from cocoa beans, water and spices, and used it to treat coughs and fever, amongst other things.

The village of Tanah ('Our Home'), in the Cayo district of Belize, is now the home of 1 500 Maya people. The surrounding countryside has many magnificent temples from the Ancient Maya and they still use the caves for ceremonies and rituals.

The Chiquibul River has its source in the hills of Belize, and flows west into Guatemala, joining other streams before ending in the Caribbean Sea. The river mainly flows underground, through limestone. The power of the water has **eroded** some of the largest caves in the world.

Since the 1960s, over 300 caves and 150 miles of passages have been discovered in the area. The most spectacular cave system in Belize is named after the river, the Chiquibul. It is found to the west of the Maya Mountains on the border between Belize and Guatemala. It is thought to contain the longest underground passage in Belize and Central America.

The Chiquibul cave system includes the Cebada cave, which is the largest in Belize and the fourth largest in the world. On the surface is the feature known as the Natural **Arch**, a huge limestone arch created by the Chiquibul River.

△ **Figure 5.21** Chiquibul cave system in Belize

△ **Figure 5.22** Stalactites of calcium carbonate hang from the cave roof like icicles

Activities

1 Use Figure 5.21, an atlas and the websites to produce your own location map for the Chiquibul cave system. Try to include the following labels:

- Caribbean Sea
- Border with Guatemala
- Maya mountains
- Chiquibul Forest
- Chiquibul River
- Chiquibul cave system

Chiquibul National Park

The Chiquibul cave system is within a National Park. You may remember that Winnat's Pass is in the Peak National Park and Malham Cove is in the Yorkshire Dales National Park.

There are many different activities available for **tourists** visiting the caves, including hiking, wading through water, tubing (floating in an inner tube), canoeing and climbing.

Many caves prove to be a challenge for even experienced **potholers** or 'spelunkers'. There are a few caves beginners can visit, for example the Rio Frio Cave. Caves are dangerous environments and visitors to the caves should always have the correct equipment with them.

The Tourist Board recommends five essential items for potholers:
- headlamps (and spare batteries)
- sensible shoes
- drinking water
- light snacks
- an experienced guide

❭ Stalactites and stalagmites.

Most of the caves have **stalactite** and **stalagmite** formations and some have spectacular waterfalls. There is evidence that the caves have been used by people for a long time. The remains of pottery and human skeletons from the Ancient Maya people have been found in some of them.

▷ **Figure 5.23**

Safety issues and helpful tips for cave visitors:

If you are **claustrophobic**, know your limits. Rio Frio is an open cave but most caves tend to have no natural lighting permeating.

Caving can be dangerous with many caves prone to **flash floods** during extended rainy periods.

Ask about the lighting being used on your tour. An extra flashlight (preferably waterproof) with spare batteries is always a good idea.

Know the accessibility and physical endurance required for each cave before your journey.

Use experienced guides for all remote caves and even well travelled caves.

Pack lightly especially for the wilder, lesser-known caves where you have to travel for long distances.

Activities

Foundation

2 Produce a poster showing the five things recommended by the Belize Tourist Board for potholers.

Target

3 Using Figure 5.23, produce a poster of words and pictures to illustrate the six safety issues for visitors to the caves and the five essential items recommended by the Tourist Board.

Extension

4 Many visitors to the caves may not speak/ read English very well. Produce a poster using only pictures to illustrate the six safety issues and the five essential items recommended by the Tourist Board.

Case Study – Postojna, an example of karst scenery in Slovenia

▷ **Figure 5.24** Christmas crib, Postojna caves

A train through a cave?

The Postojna Caves in Solvenia, have a year round temperature of 8° C and extend for 20 km underground.

The caves have been used by people for hundreds of years and for a variety of reasons (see box below). It is estimated that almost 30 million tourists have visited the caves during that time. In the past, visitors were carried round in sedan chairs.

PAST AND PRESENT USES OF THE CAVES

- Scientific research
- Medieval castle built into cave entrance
- Army storage of aeroplane fuel during World War 2
- Concerts – up to 10 000 in the Concert Hall
- Tourists – by train
- Post Office

You do not need any specialist equipment to visit the caves. The caves have had an electricity supply for many years and lighting makes it easy to see the stalactites and stalagmites. The light bulbs also allow green plants to grow in the caves, using the artificial light for **photosynthesis**.

Tourists are well catered for when they visit the Postojna caves. There are plenty of parking spaces for the cars and coaches that visit from countries such as Italy and Germany. There is also a café, toilets and a gift shop.

At the start of each new tour, tour guides hold up flags, and visitors select a guide depending on their language. After a short walk and a brief introduction to the caves, visitors board a train. The open-air carriages of the train whisk people through the cave systems but they need to watch their heads on the stalactites!

△ **Figure 5.25** Location map of caves in Slovenia

Activities

1 Choose one of the uses for the Postojna caves from the box and explain why a cave would be a good location for this activity.

2 List five other possible uses for a cave.

△ **Figure 5.26** The train through the Postojna Caves

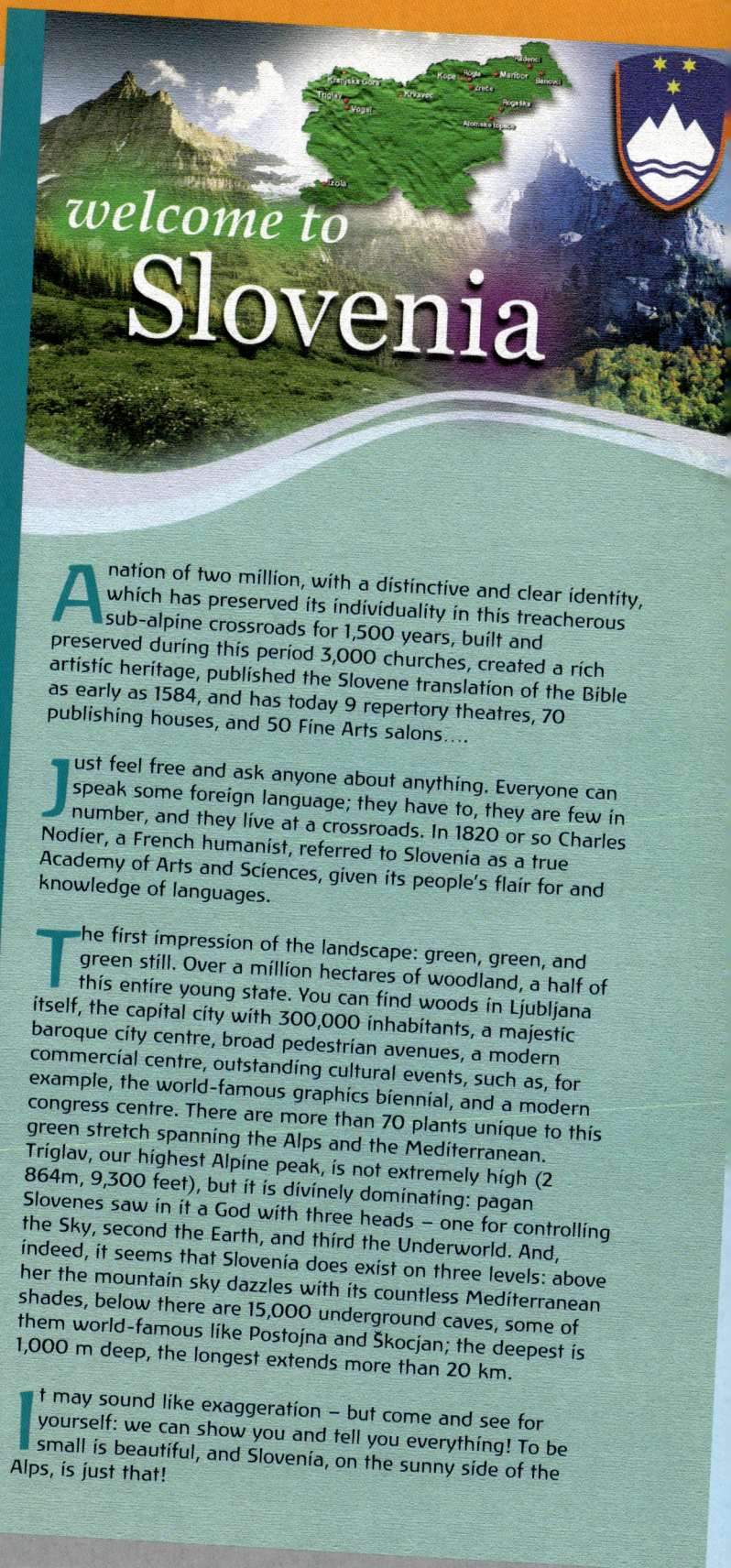

The visitors are guaranteed to see Proteus during a visit to the Postojna caves because the mysterious ruler of the deep is kept in viewing tanks. There are also over 190 other varieties of animal life in the caves including leeches, beetles, fish, shrimps, water bugs and cave louse.

Activities

3 Design a flow diagram to show how the number of visitors and the types of facilities at the Postojna caves could damage the cave environment.

4 Discuss with your partner: Why do caves like Postojna in Slovenia and White Scar Caves in Yorkshire have a constant air temperature of 8°C all year round?

5 Using the information from Figure 5.27, create a spider diagram or bullet point list to show the main attractions in Slovenia.

ICT links

You may like to find out more about the Postojna caves:

www.arctur.si/slovenia/postojna.html

welcome to Slovenia

A nation of two million, with a distinctive and clear identity, which has preserved its individuality in this treacherous sub-alpine crossroads for 1,500 years, built and preserved during this period 3,000 churches, created a rich artistic heritage, published the Slovene translation of the Bible as early as 1584, and has today 9 repertory theatres, 70 publishing houses, and 50 Fine Arts salons….

Just feel free and ask anyone about anything. Everyone can speak some foreign language; they have to, they are few in number, and they live at a crossroads. In 1820 or so Charles Nodier, a French humanist, referred to Slovenia as a true Academy of Arts and Sciences, given its people's flair for and knowledge of languages.

The first impression of the landscape: green, green, and green still. Over a million hectares of woodland, a half of this entire young state. You can find woods in Ljubljana itself, the capital city with 300,000 inhabitants, a majestic baroque city centre, broad pedestrian avenues, a modern commercial centre, outstanding cultural events, such as, for example, the world-famous graphics biennial, and a modern congress centre. There are more than 70 plants unique to this green stretch spanning the Alps and the Mediterranean. Triglav, our highest Alpine peak, is not extremely high (2 864m, 9,300 feet), but it is divinely dominating: pagan Slovenes saw in it a God with three heads – one for controlling the Sky, second the Earth, and third the Underworld. And, indeed, it seems that Slovenia does exist on three levels: above her the mountain sky dazzles with its countless Mediterranean shades, below there are 15,000 underground caves, some of them world-famous like Postojna and Škocjan; the deepest is 1,000 m deep, the longest extends more than 20 km.

It may sound like exaggeration – but come and see for yourself: we can show you and tell you everything! To be small is beautiful, and Slovenia, on the sunny side of the Alps, is just that!

△ **Figure 5.27** Extract from Slovenian Tourist brochure, showing how caves are used to attact tourism

Case Study – the White Peak, England

Look back at page 88; which describe the four main types of limestone found in Britain. Carboniferous limestone is found in the White Peak area of the Peak National Park. The area takes its name from the colour of the limestone.

Activities like mountain biking, hang-gliding, potholing and walking are very popular with people spending weekends and holidays in the White Peak. The Peak Park receives over 30 million visitors a year; many are from nearby cities such as Sheffield and Manchester. So many visitors can create conflict with residents, environmentalists and the Peak Park Authority.

Winnats Pass (Figure 5.20) is an area of outstanding beauty within the White Peak. **Accessibility** and the fantastic scenery have made the Winnats Pass a **honeypot** site. Visitors use the area for energetic outdoor activities as well as quiet relaxation. Without careful management these activities could damage the environment, for example through footpath erosion, litter being dropped, vegetation being trampled, etc. Most of the visitors travel by car and this means that pollution levels can sometimes be higher than in the centre of London.

▽ **Figure 5.29** The Houses of Parliament and St Paul's Cathedral

The caves near Castleton and Winnat's Pass have been changed by people who mined a unique type of **fluorspar**, the Blue John, found within the limestone. Today, Blue John is used to make ornaments and jewellery because of its unusual colouring.

△ **Figure 5.28** Blue John

Like many caves in the world there are tales of witches and ghosts to attract the visitors. Sir Arthur Conan Doyle (author of the Sherlock Holmes stories) wrote a short story about the Blue John caves called 'The Terror of Blue John Gap'.

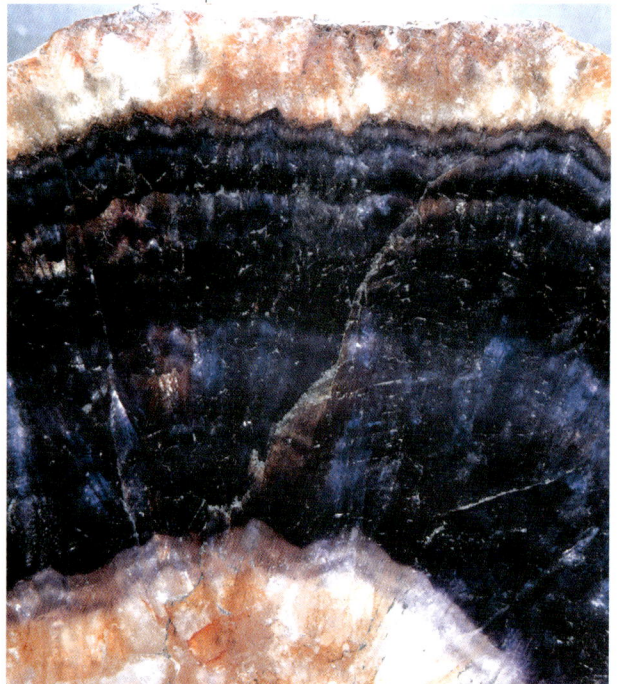

Activities

1 What is a honeypot site?

2 How do you think honeypot sites got their name?

3 Why do the visitors cause conflict with landowners, residents and environmentalists?

Limestone is used in the construction industry, such as for buildings, roads and airport runways. Some very famous buildings in London use **Portland limestone**, see Figure 5.29. Some of the more unusual uses include toothpaste, talcum powder and bread.

▷ **Figure 5.30** A typical British 'rock house' of the 21st century. We may not live in caves, but our homes are made from natural resources

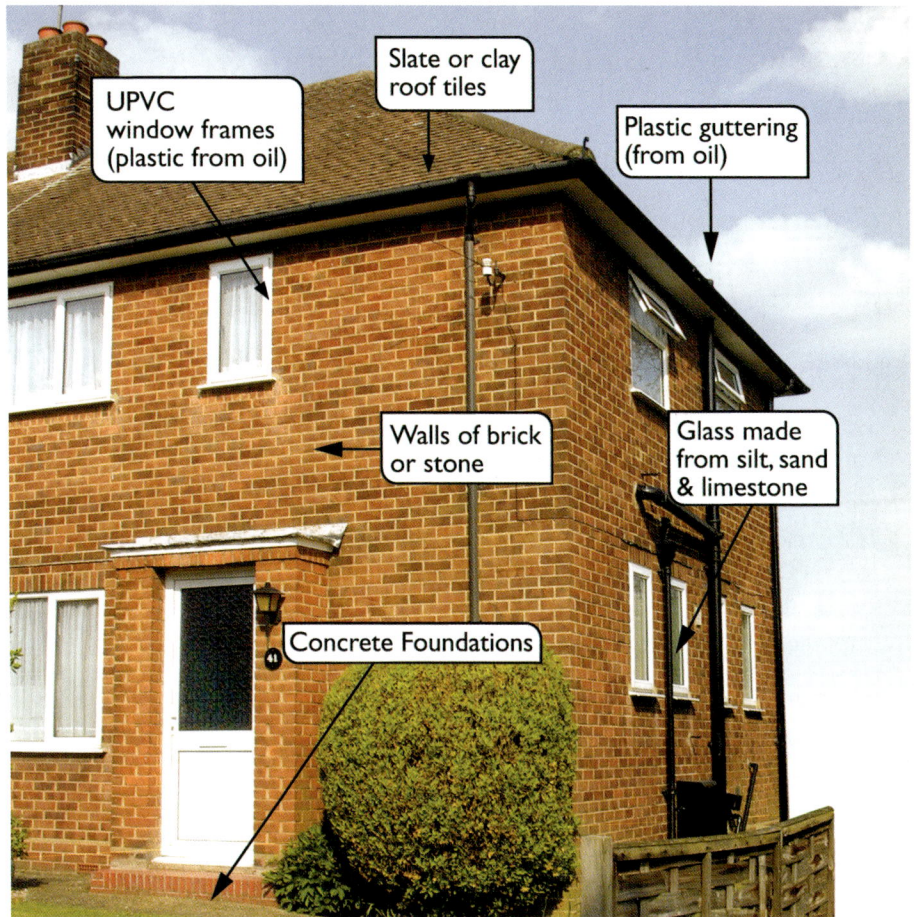

UPVC window frames (plastic from oil)

Slate or clay roof tiles

Plastic guttering (from oil)

Walls of brick or stone

Glass made from silt, sand & limestone

Concrete Foundations

〉 Conflicts over quarrying

However, before limestone can be used it has to be extracted. **Quarries** and cement works can create conflict with tourists, local residents, environmentalists and the National Park Authority because they can create pollution, reduce house prices and increase heavy traffic along narrow, country lanes. However, those people who work at the quarry, transport the limestone or use the limestone in their work will be in favour of the quarry.

△ **Figure 5.31** One of Tarmac's Quarries

△ **Figure 5.32** Philosophers Stone, Tout Quarry, Isle of Portland

Environmental issues and quarrying

Limestone is a **non-renewable resource** because once it has been **extracted** and used it cannot be replaced. The scenery it helped to create is also lost. The protester shown opposite was campaigning against a quarry expansion near Bristol. Protesters say the extension will destroy rare orchids, and disturb badgers and bats.

Quarrying creates noise (blasting with explosives, or using machinery) and dust in the air from small particles of rock. Transporting the rock to the customer can also create problems. Over 90% of the rock from the Yorkshire Dales National Park is transported by lorries. Road transport increases noise and air pollution and makes village roads busier.

Restoring the land after quarrying has finished needs careful planning but some have been successfully redeveloped as lakes, picnic areas, climbing centres and even works of art (see Figure 5.32).

The Beit Guvrin quarry and cave system is now a major tourist attraction in Israel (see Figure 5.34), even virtual tours are available on the Internet! Look at: http://israeliculture.about.com/culture/israeliculture/library/gallery/blzzbg00.htm

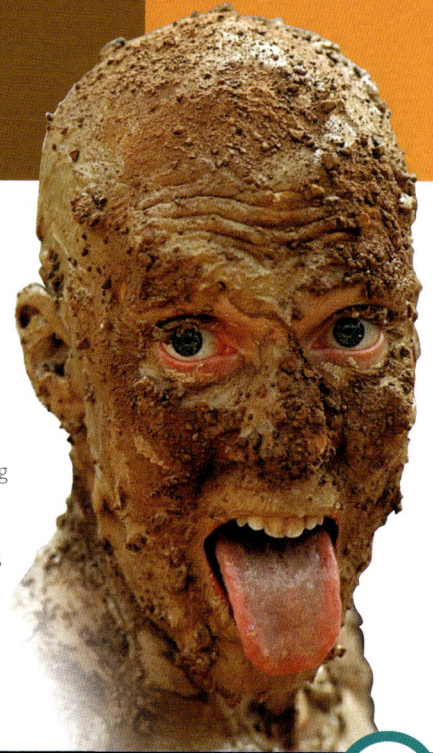

▷ **Figure 5.33**
This protester was campaigning against the expansion of Dunford quarry, near Bristol

Activities

1 Draw a spider diagram to show the environmental issues created by quarrying limestone.

2 Produce a local newspaper article about the proposed quarry extension near Bristol (Durnford Quarry at Long Ashton). Include the views of people in favour and against the proposal. You could include maps and diagrams too.

▽ **Figure 5.34**

Assessment tasks

The pictures below show some still frames from a computer game. When computer games and films are designed, a storyboard is used to plan out ideas. The storyboard is a sequence of drawings that show how the game fits together.

>YOUR TASK IS TO
DESIGN A COMPUTER
GAME ...

△ **Figure 5.35**

Target tasks

Your task is to design a computer game called 'Proteus'

1 **a** Design a character for your game to represent Proteus. Draw the character and add information about it such as what it eats, how it moves, its size, etc.

b Design the environment that the character will travel through. Include as many LIMESTONE features as you can.

c Write 20 tricky questions with three possible answers for each, to test the player on limestone. Include ten challenge spots where the player of the game will need to test their knowledge of limestone.

Assessment tasks

Extension tasks

2 Your main character, Proteus, must negotiate a course through limestone scenery, from a swallow hole and limestone pavement on the surface, through a cave network, to the emerging spring water.

As your character travels through this subterranean world, they will move vertically and horizontally through the cave system. To change level and direction, you will need to devise tasks, obstacles or questions that test the player's knowledge of limestone scenery and the human activities that threaten this underground world.

Other things to consider:
1 You will need to plan out a cross-section of the limestone scenery.
2 You must design the character of Proteus: how it moves, how it eats, weaknesses, etc.
3 Are there realistic threats to the environment, such as human activity in the caves, and where might this appear in the cross-section?
4 Are there other characters in the game, or obstacles to overcome?
5 You need to plan the questions (and answers) you will ask along the way. Will they be more difficult as you progress further in the game?
6 What is the aim of the game? How will you decide when it is over? Will there be a winner? Is this a game of survival?

Review

Test your knowledge on limestone

CAN YOU NAME THREE...

types of rock?

sedimentary rocks?

types of limestone?

periods of geological time when limestone was created?

countries where limestone is found?

area of Britain with limestone?

uses of limestone?

tests to identify limestone from other rocks?

interesting facts about Proteus?

ways in which the rock type can affect the landscape?

creatures that have helped to make limestone?

safety tips for potholers?

uses for caves?

threats to limestone?

management strategies?

△ **Figure 5.36**

Activities

1 Copy out the cross section in Figure 5.36 and label as many features as you can.

Where there is muck ...

▽ **Figure 6.1a**

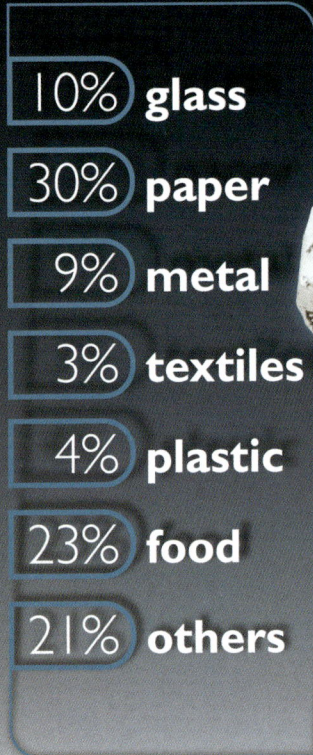

- 10% **glass**
- 30% **paper**
- 9% **metal**
- 3% **textiles**
- 4% **plastic**
- 23% **food**
- 21% **others**

△ **Figure 6.1b**

... there's brass!

Activities

1 What do Figures 6.1a and 6.1b say about how we use our resources?

a What do we throw away? Why?

b Where does our waste end up?

c Why do children search rubbish tips? Would you?

d Is our rubbish worth anything?

△ **Figure 6.2** Children collecting from an LEDC tip in Dhaka, Bangladesh

What about us?

In MEDCs we live in a consumer society. We buy things, use them for a while, and then throw them away. Industry manufactures goods cheaply. Advertisers persuade us to buy things.

Twenty years ago the average car was ten years old; today it is four.
So what do we use? How long did we have it before we disposed of it?

▽ **Figure 6.3** What goes in and out

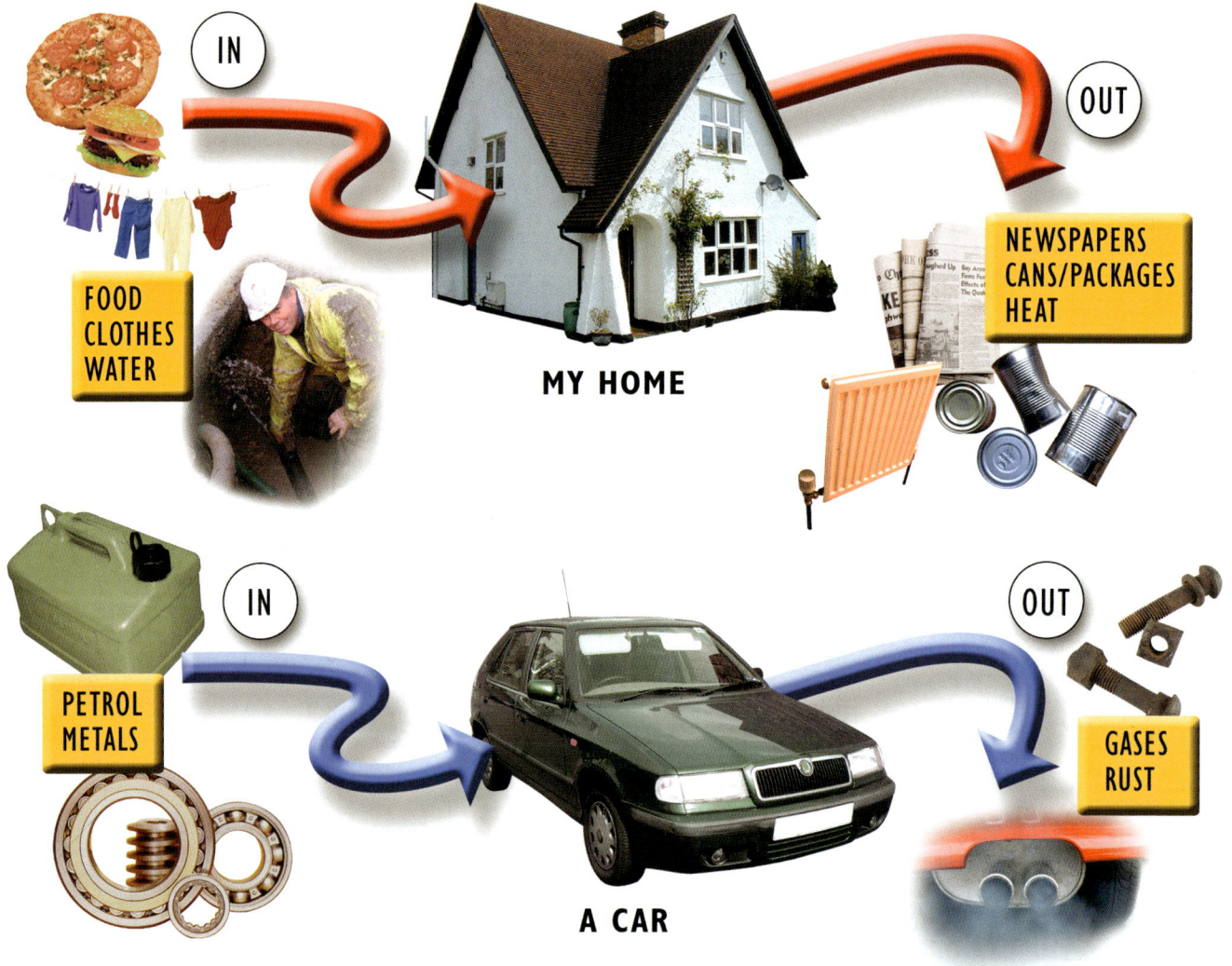

IN

FOOD
CLOTHES
WATER

MY HOME

OUT

NEWSPAPERS
CANS/PACKAGES
HEAT

IN

PETROL
METALS

A CAR

OUT

GASES
RUST

Activities

1 **a** Look at the two central pictures. List ten things that go into each, and suggest ten waste items.

b Look at your home shopping list: which items do you think are essential? Why?

c How much waste does your household produce each week?

d Where does this waste go?

e What happens to this waste?

What if?

▽ **Figure 6.4**
Uses of oil

What if the world runs out of oil in the next fifty years? How would it affect us? Oil is a very important raw material. It supplies 40% of the world's energy. It has many uses in the chemical industry and in manufacturing (Figure 6.4).

Oil is the microscopic, compressed remains of sea creatures which lived millions of years ago. It is found underground trapped beneath layers of impermeable rock. Oil companies extract the oil by drilling through the rock. Most of the accessible deposits have been exhausted. Today oil is being extracted from hostile environments like the Arctic or North Sea (Figure 6.5). No one knows how much oil is left, but once oil has been used it cannot be replaced.

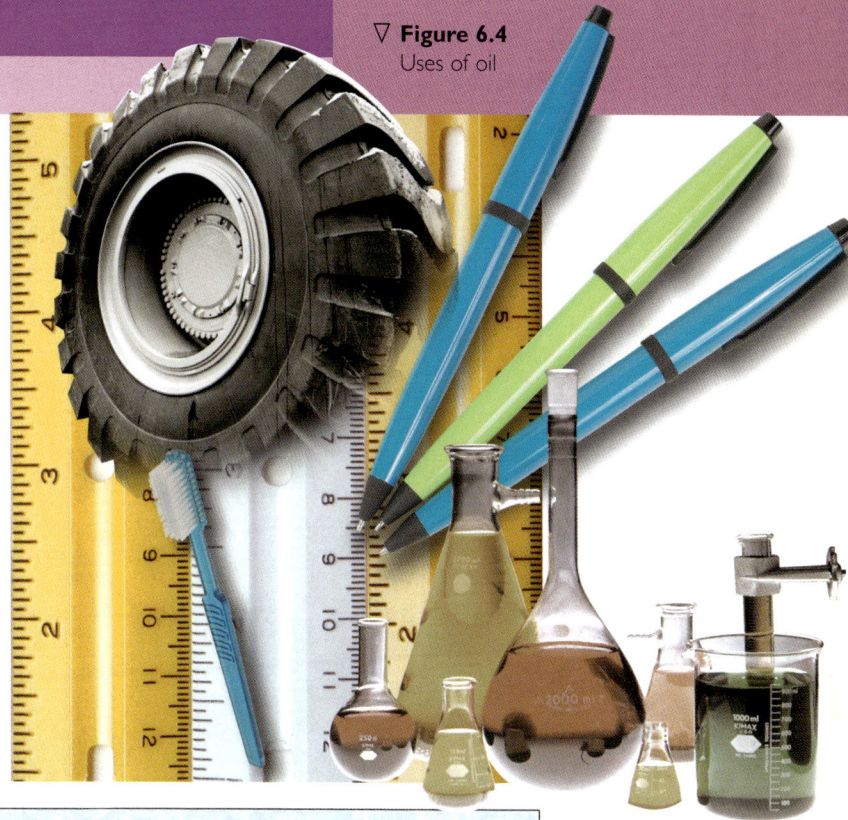

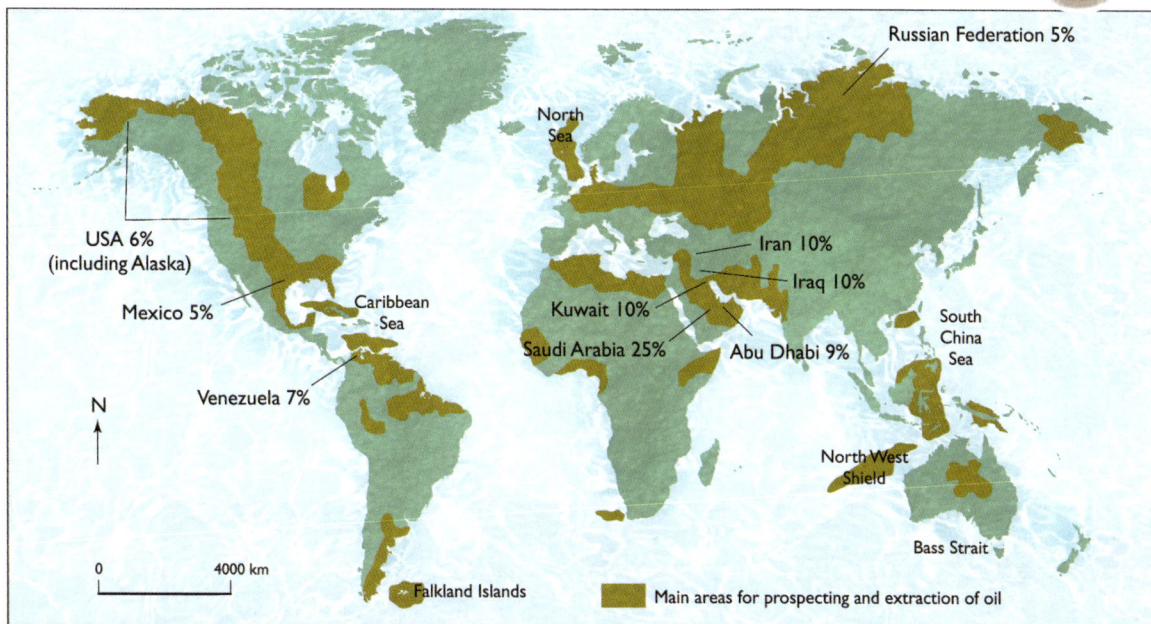

△ **Figure 6.5** Countries with more than 2% of the world's oil deposits

Activities

2 Imagine that the world's oil supplies do become exhausted. With a partner, use a concept map to describe how this would affect our lifestyle.

3 Suggest five other resources that you think will become scarce in the next ten years. How might this change our lives?

4 Use Figure 6.5 to identify where the world's oil reserves are located. Suggest why this may be a problem in the future for MEDCs.

Spaceship Earth

ALL RIGHT, WHO USED THE LAST DROP OF OIL?

△ **Figure 6.6**

Imagine that the whole of humanity is on an incredible journey. Our spaceship is planet Earth. On board we have everything to sustain a diversity of life. But there is nowhere to stop to replenish our supplies or repair the damage to the environment. If our resources run out or if the spaceship's **ecosystem** fails …

Each year 24 million tonnes of household waste is produced. That is one third of a tonne per person. Industry adds even more. All waste is potentially dangerous and can damage the environment. Some waste is recycled and re-used. This will help conserve finite resources and care for the environment.

The world's population is eight billion. It will continue to grow faster in the LEDCs compared to the MEDCs. Each person will want a share of the limited resources.

Each person will want water, food, farm land, land to build a house and fuel. These will provide some of our basic needs. We also need many other things to give us a reasonable **standard of living**. But the world's resources are not evenly distributed.

Many important resources are becoming scarce and difficult to extract. The issue is how to manage our limited resources to meet the needs of future generations.

▽ **Figure 6.7a and b** Who has access to the world's resources?

The UK only recycles 9% of its waste. However, there are many organisations which suggest how resources can be conserved and how waste can be recycled. Many have their own websites or work with local councils to set up recycling facilities. In some places it is compulsory to sort your waste before collection. A major problem is that the prices paid for sorted waste materials may not be enough to make it worthwhile.

ICT links

Recycling websites:
www.greenpeace.org.uk & **www.foe.co.uk**
These organisations deal with a range of issues.

www.wastewatch.org.uk & **www.crn.org.uk** have information on waste reduction and reuse.

www.greenstat.co.uk sell stationery made from recycled paper.

www.realnappy.com Disposable nappies make up 4% of UK landfill site material.

www.cat.org.uk The Centre for Alternative Technology has practical advice.

www.theecologist.org investigates issues caused by our consumer world.

www.greenguide.co.uk A news site and database about green issues.

http://environment.ngfl.gov.uk An attractive site which allows you to test your knowledge of the environment.

Activities

1 Plan a survey to investigate people's attitudes toward waste and recycling:
- Think of five questions
- Ask 20 people
- Collate your results
- Use an appropriate presentation for your results
- Explain what you have discovered.

2 Find out if your school recycles waste or buys products made from recycled materials, for example, paper. Write a letter to the school governors to suggest how the school could use recycled resources and produce less waste.

3 Discuss with a friend three ways the world's resources can be conserved. You could use one of the websites. Write each idea on a piece of paper. Draw out the continuum grid, Figure 6.8. Place each of the three ideas on the line where you think it best fits.

MOST PRACTICAL	10–9–8–7–6–5–4–3–2–1–0	LEAST PRACTICAL
GREATEST IMPACT	10–9–8–7–6–5–4–3–2–1–0	LEAST IMPACT
CHEAPEST	10–9–8–7–6–5–4–3–2–1–0	MOST EXPENSIVE

△ **Figure 6.8** Continuum grid

- Which is the best of the three ideas? Explain why.
- Design an advertising campaign to convince people to adopt this idea.
- Think about: What type of advert will you use? Where could it be seen? Who is it aimed at?

I think the best idea is…

Three ideas are…

What is it made from?

This will sell because…

Who eats the best?

The poorest nations have the least food and suffer from diseases caused by a poor diet. The richest nations have the most food … and suffer from diseases caused by a poor diet.

For example, the richest nations have the most **obese** populations and the greatest amout of weight related illnesses that go with this unhealthy diet (Figure 6.9).

USA – 23% obese

Sweden – 7% obese

Finland – 14% obese

Japan – 2% obese

China – 1% obese

Germany – 24% obese

Australia – 13% obese

UK – 16% obese

Very low risk of malnourishment, not less than 5% affected

Moderately low risk of malnourishment, 5–19% of population affected

Moderately high risk of malnourishment, 20–34% of population affected

High risk of malnourishment, 35% of population affected

0   4000 km

△ **Figure 6.9** Who has enough to eat?

MEDCs have the capacity to produce more food than they need. They use high yielding, disease-resistant seeds and machines to be very efficient. They also buy from LEDCs those products they cannot produce. LEDCs sell food products to earn money even though their populations may need this food or need the land to grow staple crops, such as wheat or maize.

MEDCs have a diet high in protein, carbohydrates and fats. The food contains the vitamins needed for health. This diet is more than the 2 400 calorie intake which is the estimated requirement to avoid malnutrition. LEDC diets lack protein and fats, and there are more carbohydrates but less vitamins. The overall number of calories is fewer.

Country	% of the population who are undernourished	% who are farmers	% of income from farming	GNP/ Person $
Burundi	63	92	55	160
Eritrea	67	93	80	100
Ethiopia	51	88	60	100
Kenya	41	81	29	280
Rwanda	37	90	41	180
Somalia	73	76	65	500
Sudan	20	72	34	750
Tanzania	40	85	56	120
Uganda	28	86	53	240

There is enough food in the world to feed everyone. However, 800 million people daily experience hunger and a poor diet. Children suffer the most. The lack of food causes malnutrition and can affect mental and physical development. The UN estimates that one child dies every seven seconds from a poor diet.

Hunger is related to poverty. Poor people cannot buy enough food. Others do not have the land, seeds or tools to grow crops. When **subsistence farmers** have their crops affected by drought, pests and war they have nothing to fall back on.

Many organisations help to prevent hunger. Famines are tackled by bringing in large amounts of food. But reducing poverty by making farming more productive, and by teaching skills, are potentially more helpful.

△ **Figure 6.10** East Africa indicators

Activities

Foundation

1 Which parts of the world have a poor diet?

2 Why is a poor diet a problem?

3 Suggest two reasons why malnutrition occurs in poorer countries.

4 How could diets in LEDCs be improved?

Target

5 Present the data in the table above (Figure 6.10) as four maps, using an appropriate method.

6 How do your maps show that East Africa is a region that might suffer food shortages?

7 Suggest why there is a link between low income and health.

Extension

8 If a person from Ethiopia came to the UK, what would s/he find out about our eating habits?

9 Why might what they found surprise her/him?

Farming systems

In the UK only 2% of the workforce is employed in farming. UK farms are **agribusinesses**. In LEDCs many families are subsistence farmers who use simple technology to grow and supply their food (Figures 6.11a and b). UK farmers can rely on help if their crops fail. The UK can import food if it needs to. Subsistence farmers face ruin and famine if their crops fail. The fragile farming system means that LEDC food production finds it difficult to keep up with a growing population.

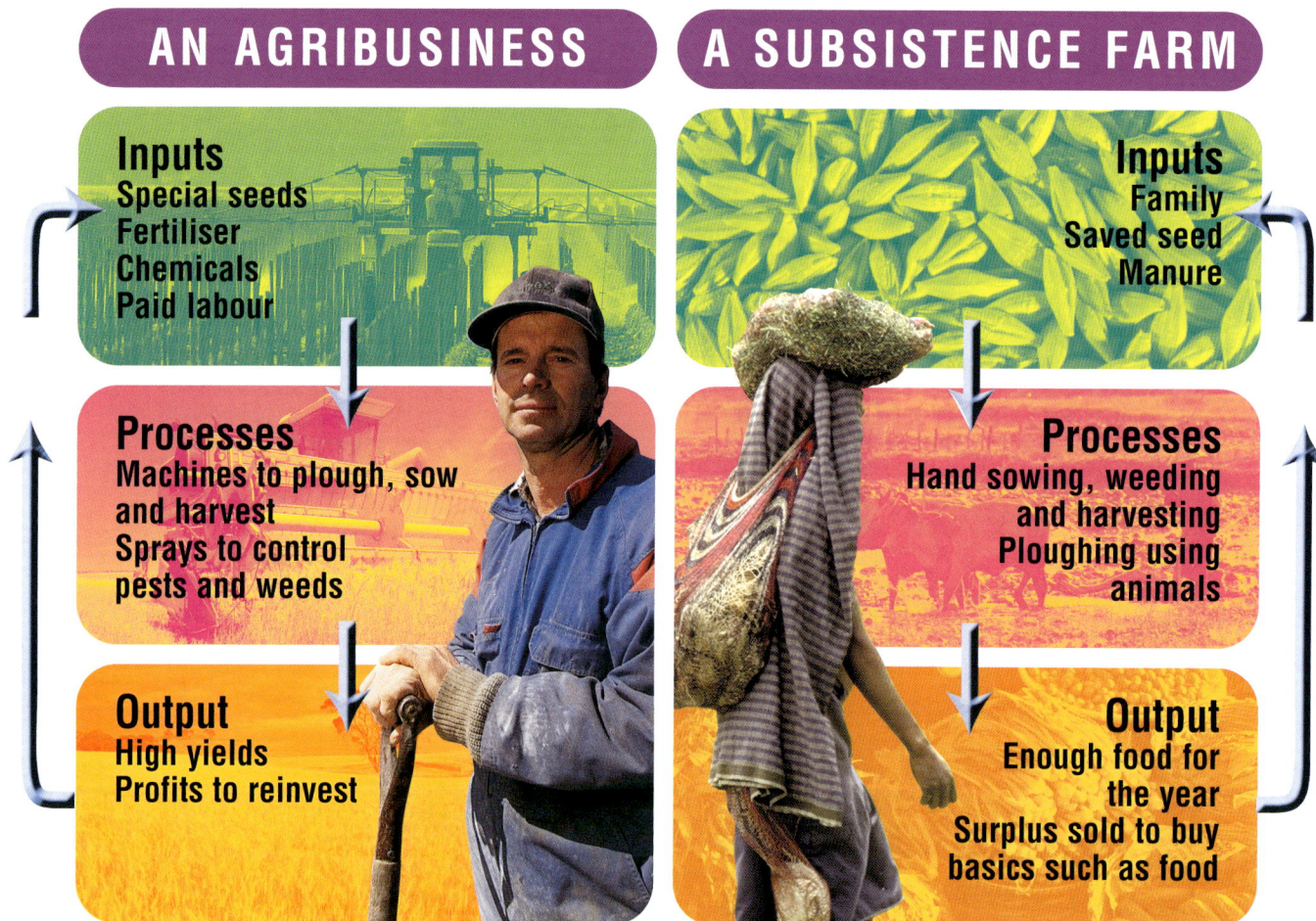

AN AGRIBUSINESS

Inputs
Special seeds
Fertiliser
Chemicals
Paid labour

Processes
Machines to plough, sow and harvest
Sprays to control pests and weeds

Output
High yields
Profits to reinvest

△ **Figure 6.11a** An agribusiness in an MEDC

A SUBSISTENCE FARM

Inputs
Family
Saved seed
Manure

Processes
Hand sowing, weeding and harvesting
Ploughing using animals

Output
Enough food for the year
Surplus sold to buy basics such as food

△ **Figure 6.11b** A subsistence farm in an LEDC

Technology has helped some LEDC farmers. High yielding varieties (HYVs) have increased production – but only for the farmers who can afford the fertiliser and extra water these seeds need. This approach has also caused environmental damage.

Genetically modified (GM) seeds are seen by some as a way to improve the reliability of the world's food supply. However, others think that pollen from GM crops will damage natural plants and threaten the environment.

Intermediate technology may be a better way to grow more crops and look after the environment. Simple ideas that use local materials and skills can make the soil more productive. For example, breaking the slope with lines of stones stops soil being washed away but allows more water to soak into the soil.

Fresh fruit and vegetables all year

Whatever the month, supermarkets stock a wide range of fruit and vegetables even when they are out of season in the UK. They can supply their shops with fresh produce grown thousands of miles away in the tropics. The all-year warm climate, good soils and irrigation help to ensure a guaranteed supply. Cheap air-freight and electronic data exchange (EDI) systems help order and deliver produce within 24 hours.

Growers in southern Africa plant seeds every four days to meet the anticipated demand. The EDI link helps to monitor sales and uses weather forecasts to predict demand. As soon as an order is placed, the vegetables are picked, sorted and packed by a large number of people. Most of the workers are women, who are paid an above average wage and provided with lodging and food. A kilo of vegetables is sold for £1. In the UK this kilo retails for £9.

The industrialised nations have changed their eating habits to take advantage of this supply. However some believe it is wrong to exploit LEDCs in this way. Water resources and land are being used to earn money rather than feed local people. Others warn that too much pesticide is used, which can harm the workers.

deliveries to supermarkets

aircraft lands and vegetables are taken to a central depot for redistribution

supermarkets order their next day's vegetables

order received in Zimbabwe

Zimbabwe

aircraft leaves for the UK

vegetables are picked and packed by hand

taken by lorry to the airport

△ **Figure 6.12** Supplying UK supermarkets

Activities

1 List three differences and three similarities between an agribusiness and a subsistence farm.

2 What are the arguments for and against LEDCs providing MEDCs with fresh fruit and vegetables all year round? Set your work out as a table.

3 Visit the National Farmers' Union website at www.nfu.org.uk. This site has farming case studies from around the world. Choose two contrasting examples and draw an input–process–output diagram for each of them. (Inputs are what goes in, processes are what happens on the farm, and outputs are what is produced.)

4 Below is a start to a 'Most likely to…' exercise. Write 8 more statements of your own, making sure some of them lead to a discussion.

Which farm is most likely to…
have a tractor? buy seed in bulk? export most of its produce?

Now share your statements with the class.

How is soil a resource?

Farming is the key to producing food to feed the world's population. The gases in the atmosphere which all living things need to survive depend on plants carrying out photosynthesis and keeping the gases in balance. Both of these processes depend on the earth's thin layer of soil.

So, soil is a vital resource for all life on earth. But, the fertility of large areas of soil is under threat. In the last 50 years about 15% of land has lost its fertile soil. Some areas may never grow plants again. The change from productive land to desert is called desertification.

Some of these changes can be blamed on climatic change, but much has been caused by human activity, particularly in LEDCs where people are trying ever harder to feed increasing populations from subsistence farming. Often the time is not available to allow the soil to recover before planting another crop. People also need fuel and often chop trees down for firewood. Too many animals are allowed to graze on an area. This all means that the soil does not have its nutrients replaced. It is exposed to the weather and might be blown by winds or washed away by heavy rain.

▽ **Figure 6.13** World map showing desertification by grade

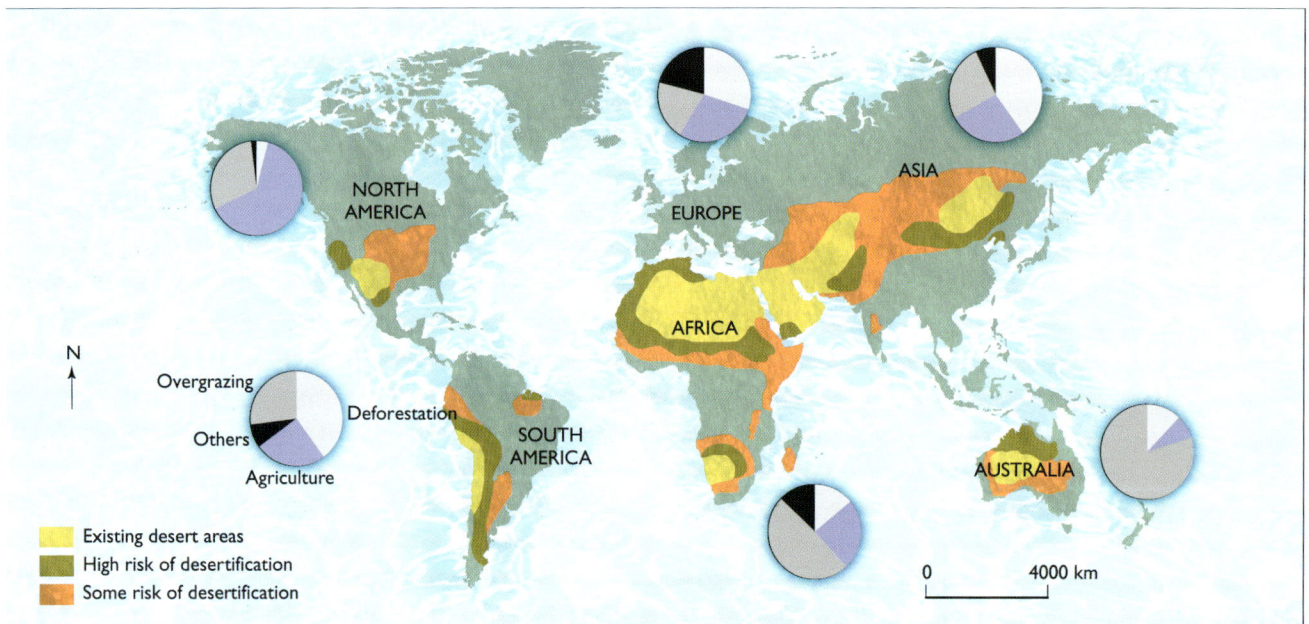

Soil is a limited resource. It needs to be looked after carefully if it is to continue providing crops year after year. Whilst there is a problem in MEDCs too, the problem is life and death for many in LEDCs. Reversing desertification is essential if rural populations in the main affected areas are to survive. Education is an important part of the process. Aid agencies work with local villages to show them how to look after their soils, using simple methods and technologies.

◁ **Figure 6.14** Soil erosion

1 Feed an increasing population	2 Overgrazing	3 End of traditional **Nomadic** lifestyles
4 Afforestation – plant trees to cover soil and act as windbreak	5 Add animal manure on to land	6 Climate change partly caused by **fossil fuel** use
7 Provide aid from MEDC	8 Keep fewer animals	9 Provide alternative fuels in both LEDC and MEDC
10 Firewood needed for cooking	11 Ploughing up and down the slope	12 Ploughing around the slope – following the contours
13 Increasing rural population	14 Deforestation	15 Poor farming methods
16 Encouraged to use manufactured fertiliser rather than manure	17 Provide better education	18 Leave some land **fallow** each year
19 Work to reduce the greenhouse effect	20 More land being used to grow crops for export	21 Wildlife conservation areas created

△ **Figure 6.15** Causes, reasons and solutions of soil erosion

Activities

1 Use the map (Figure 6.13) to write a description of the **distribution** of the soil erosion problem around the world.

2 **a** Read the statements in Figure 6.15 and divide them into groups:

- farming methods causing soil erosion
- reasons why soil erosion is happening
- possible solutions to soil erosion.

b Select the statements that you think could best be used to annotate the photograph (Figure 6.14)?

c Arrange the statements into groups to link together causes, reasons and solutions.

3 Imagine you are an aid worker in an area with a soil erosion problem such as the Sahel. Create a storyboard to explain the causes of soil erosion and possible solutions to the local farmers.

Energy is a basic need and one of the world's important resources. MEDCs achieved industrial growth by using vast amounts of fossil fuels. These are non-renewable resources; for example, coal, oil and natural gas took millions of years to form and once used cannot be replaced. The

MEDCs use 82% of world energy production each year but only have 20% of the world's population. The USA uses 100 times more energy than many LEDCs. LEDCs use little energy use because of the type of fuel used and lack of industry, Figure 6.16.

World share of fuels	Traditional fuels	HEP	Coal	Natural gas	Nuclear	Oil
MEDCs	5%	62%	65%	85%	98%	73%
LEDCs	95%	38%	35%	15%	2%	27%

△ **Figure 6.16** Comparing sources of world energy

Share of fuels	Traditional fuels	HEP	Coal	Natural gas	Nuclear	Oil
Nigeria	62	2	0	8	0	28
UK	0	1	32	24	7	35

△ **Figure 6.17** Energy profile for the UK and Nigeria

While MEDCs look for ways to secure future energy supplies, many LEDCs are struggling to meet the demand for energy. Their lack of an infrastructure means that electricity generation and distribution is not possible. The high population growth rates will increase the demand for energy. The energy options left for the rural and poor are fuelwood, charcoal, kerosene and dung.

In rural India cattle dung is an important resource. It is used as a building plaster, a fertiliser and as a cooking fuel. As trees have been cleared for crops, there is less fuelwood. Cattle owned by men provide the dung for the women to collect and process. Dung provides 30% of rural energy. It is estimated that 400 million tonnes of dung are burnt annually in Asia and Africa. If it were used as fertiliser it would increase grain production.

▽ **Figure 6.18** World consumption of traditional fuels

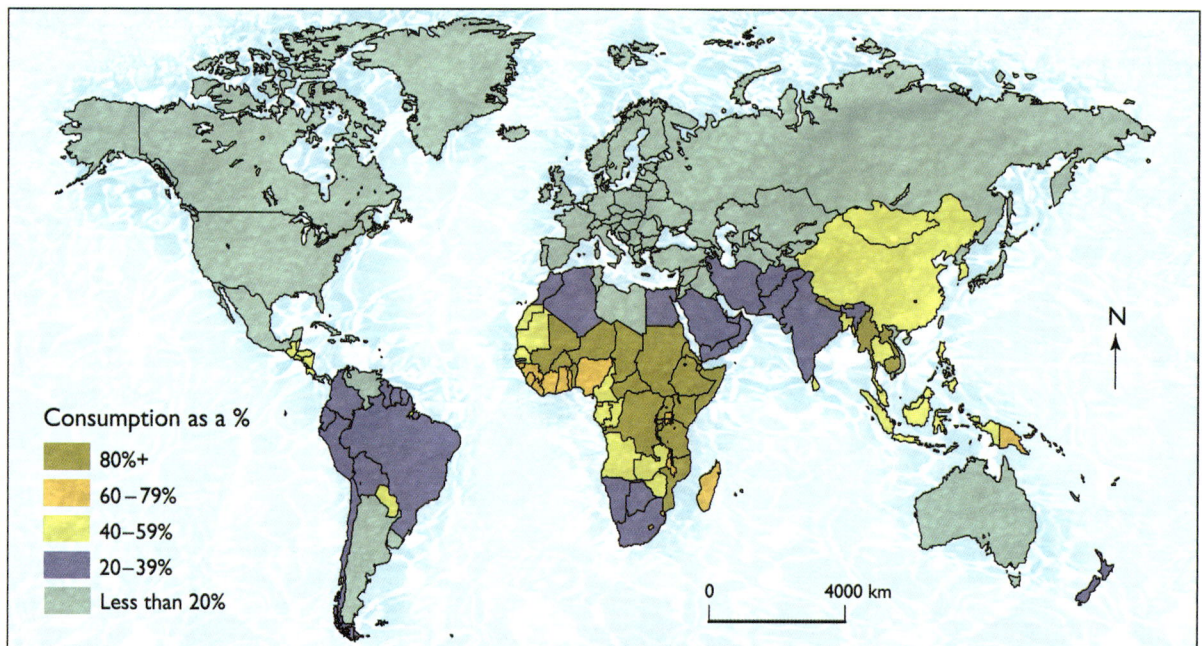

Consumption as a %
- 80%+
- 60–79%
- 40–59%
- 20–39%
- Less than 20%

0   4000 km

N

Energy for all?

Nearly half the world's population relies on **biomass** fuels to meet basic needs such as cooking and heating. Biomass fuels are burnt inefficiently, cause deforestation and pose health problems through exposure to smoke.

The use of fuelwood in Africa's Sahel has contributed to 10% of the region becoming desert. As the population grows, more vegetation is collected. As wood becomes scarce, people travel further to collect it or it becomes expensive to buy. However, the land surrounding rural settlements is cleared and left exposed. This cleared land soon becomes a desert. People living in the slums of African cities like Nairobi go into the countryside to collect wood. Other fuels are too expensive to use. It is suggested that a 1% increase in the urban population will result in a 14% increase in the use of fuelwood and charcoal.

Country	1970			1995		
	Energy use per person	Population	Urban population	Energy use per person	Population	Urban population
Chad	0.02	3.6 million	12%	0.04	6.7 million	22%
Ethiopia	0.03	28 million	9%	0.07	58 million	13%
UK	4.8	55 million	89%	5.4	58 million	90%
USA	10.3	205 million	74%	10.75	263 million	77%

△ **Figure 6.19** Comparing energy change, 1970 and 1995

A cheap intermediate technology solution to improve the energy efficiency of fuelwood involved compacting charcoal dust using a hand press. Other ideas to reduce reliance of fuelwood are mini-hydro electric schemes, use of wind turbines to charge batteries or solar panels.

Activities

Foundation

1 Describe the differences in the types of energy sources between MEDCs and LEDCs.

2 List the different sources of energy used in your house and school. Which of these are finite sources?

Target

3 Use Figure 6.19 to identify the regions which depend on traditional energy sources.

4 Suggest why these regions rely on traditional sources.

5 Explain how traditional energy sources are used.

Extension

6 Use Figure 6.19 to predict how energy use per person will change as the urban population increases.

7 What energy problems do rural areas of LEDCs face, and how might these be overcome?

Can burning fuel cause damage?

Most of our energy comes from fossil fuels. The mining of these fuels destroys the landscape and burning releases gases and pollutants into the air. Fossil fuels are linked with pollution and changes in the global climate. These concerns have been discussed at international meetings, for example, Rio 1992 and Kyoto 1997. These meetings have predicted a temperature increase of 5°C by 2100 with devastating consequences for life (Figure 6.20).

International agreements have set targets to reduce the emission of greenhouse gases like carbon dioxide and methane which come from industry, cars and farming (Figure 6.21). The industrialised countries disagree on how to achieve these targets. They fear that reductions will affect their industries while the LEDCs will hinder their development.

Part of the solution is to develop environmentally friendly **renewable energy sources**, for example wind, wave, water and solar. These have immense potential to supply large areas but only provide 6% of the world's energy needs. **Reforestation** will help reduce carbon dioxides by acting as 'carbon sinks' and providing a renewable biomass source.

▽ **Figure 6.20** Temperature change predictions

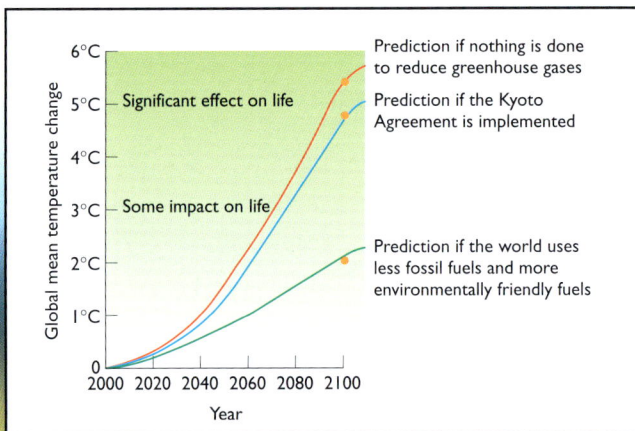

Prediction if nothing is done to reduce greenhouse gases

Prediction if the Kyoto Agreement is implemented

Prediction if the world uses less fossil fuels and more environmentally friendly fuels

▽ **Figure 6.21** World Carbon dioxide emissions

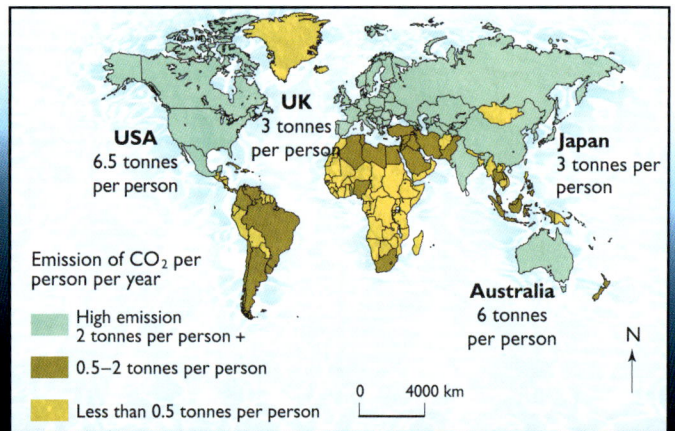

UK
3 tonnes per person

USA
6.5 tonnes per person

Japan
3 tonnes per person

Australia
6 tonnes per person

Emission of CO_2 per person per year

High emission
2 tonnes per person +

0.5–2 tonnes per person

Less than 0.5 tonnes per person

0   4000 km

N

▽ **Figure 6.22** The Greenhouse Effect

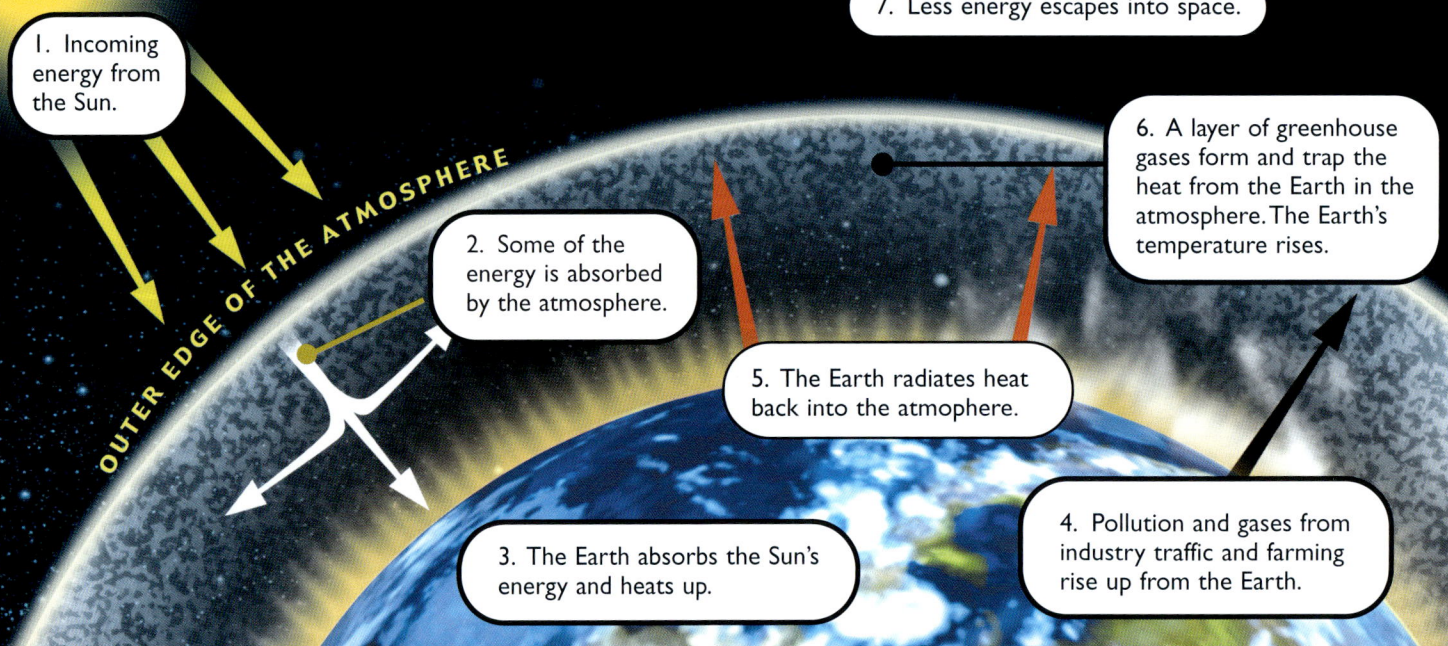

OUTER EDGE OF THE ATMOSPHERE

1. Incoming energy from the Sun.

2. Some of the energy is absorbed by the atmosphere.

3. The Earth absorbs the Sun's energy and heats up.

4. Pollution and gases from industry traffic and farming rise up from the Earth.

5. The Earth radiates heat back into the atmophere.

6. A layer of greenhouse gases form and trap the heat from the Earth in the atmosphere. The Earth's temperature rises.

7. Less energy escapes into space.

Feeling warm?
Switch the heater off

Our planet is warmed by the Sun whose heat is trapped by the atmosphere. Energy from the Sun can pass through the atmosphere to warm the Earth. Fossil fuel gases like carbon dioxide collect around the Earth like a blanket. This traps the heat being radiated from the Earth's surface, creating the Greenhouse Effect (Figure 6.22). This is causing a slow rise in the average temperature, leading to **global warming**. There is some evidence that sea levels are rising and polar ice caps are melting. As the climate changes there are predictions of extreme weather occurring. For the UK it will mean summer **droughts**, winter storms and floods, and much higher temperatures.

Rainfall is naturally acidic, and mixed with sulphur dioxide and nitrogen oxides it reacts to produce acid rain. This can kill trees, pollute rivers and chemically weather buildings. Acid rain is an international threat. The UK's fossil fuel emissions are blown by wind to fall as acid rain in Europe.

In recent years the weather forecasts have included an air quality warning. Many urban areas trap exhaust fumes which react with the higher temperatures to produce a photochemical smog. This is a low brown haze which poses a severe health hazard.

Activities

1 List at least 20 uses of electricity you benefit from on pieces of card. Decide which are:
a essential
b useful.

With a partner reduce each pile and give a reason for keeping or rejecting each card. Aim to be left with the only the most essential uses. List these and explain your final selection.

2 Explain why sea level might rise in England. In pairs draw a concept map to show how sea level rise might affect people. Think about homes, jobs, leisure, and people who live away from the coast.

▷ **Figure 6.23** Impact of sea level rise on England

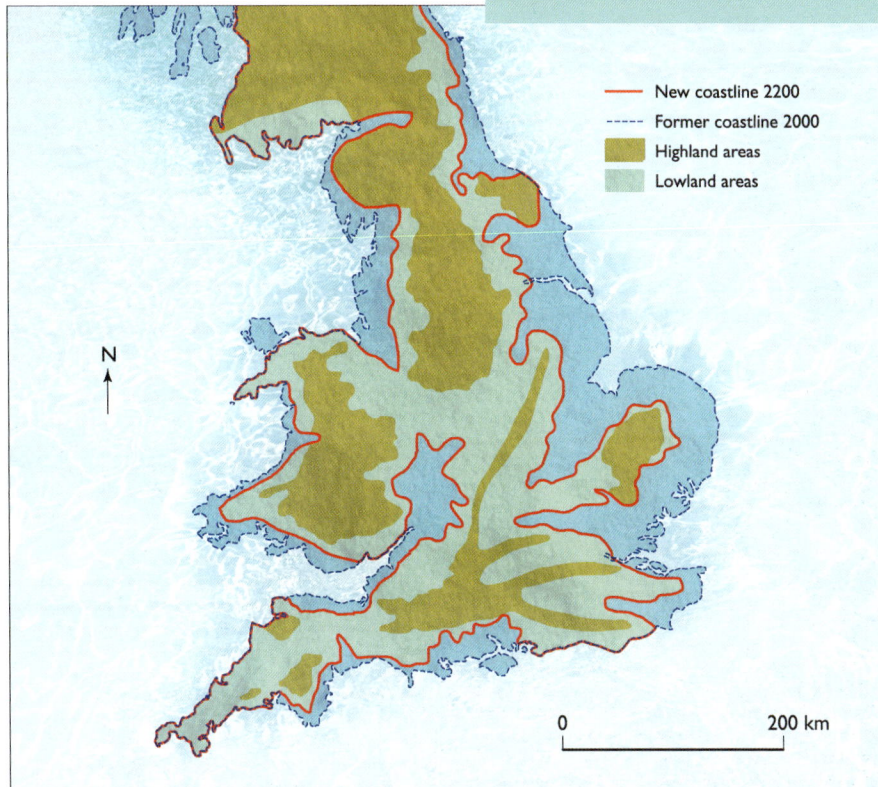

New coastline 2200
Former coastline 2000
Highland areas
Lowland areas

N

0   200 km

Assessment tasks

Background

The demand for resources will continue. MEDCs may find it easier to meet the needs of their populations. LEDCs are seeking to provide the most basic needs, especially in the rural areas.

The aim of this assessment is to consider how a sustainable energy supply can be provided and the issues that can arise.

Date	1950	1960	1970	1980	1990	2000
Amount used	3 400	3 800	5 200	6 500	10 300	14 000

△ **Figure 6.24** World Energy Consumption (all energy is based on the coal equivalent as million tonnes)

Target tasks

1
a Plot Figure 6.24 as a line graph.

b Write a sentence to describe how the world's demand for energy has changed.

c Predict the demand for energy for 2010 and 2020 and draw it onto your graph.

d Give a reason for your prediction.

2 Study Figure 6.25 (page 122). This is a location in the UK where a wind farm is to be built.

a List three factors that are important for a wind farm.

b Which of the three sites shown do you think is the best and why?

3 Study Figure 6.26 (page 122). This is in a village in an LEDC. The village has been given three wind generators. They must decide where to locate them and how to use the electricity to help meet their everyday needs.

a Make a list of five best uses of electricity for the villagers.

b On a copy of Figure 6.26, label where you would use the electricity.

c Explain how your choices will help the villagers.

4 What would you think if a wind farm was built near to where you live?

Extension tasks

1 Study Figure 6.24.

a How do you think the demand for energy will change in the next 50 years?

b Suggest two reasons for this trend.

2 Study Figure 6.26 (page 122). This is a location in the UK where a wind farm is to be built.

a Use a copy of the table below (Figure 6.25) to describe the location of each site and to note at least one advantage and problem of the sites.

b Which do you think is the best site? Explain the reasons for your choice.

c What are the advantages and disadvantages of wind power as an energy source?

3 Study Figure 6.27 (page 122). This shows a rural location in an LEDC. This village has been given a small wind farm. The village has to decide where to locate the wind turbines and how to use the electricity generated to meet everyday needs.

a Explain how the villagers are able to meet their present needs for:
  • water for domestic use and washing
  • fuel for cooking
  • irrigation for crops.

b Explain how a reliable source of electricity will help the villagers meet their future needs for:
  • water for domestic use and washing
  • fuel for cooking
  • irrigation for crops.

c What other ideas could help develop sustainable energy resources in rural areas in LEDCs?

▽ **Figure 6.25**

Site	Description of location	Advantage	Problem
1			
2			
3			

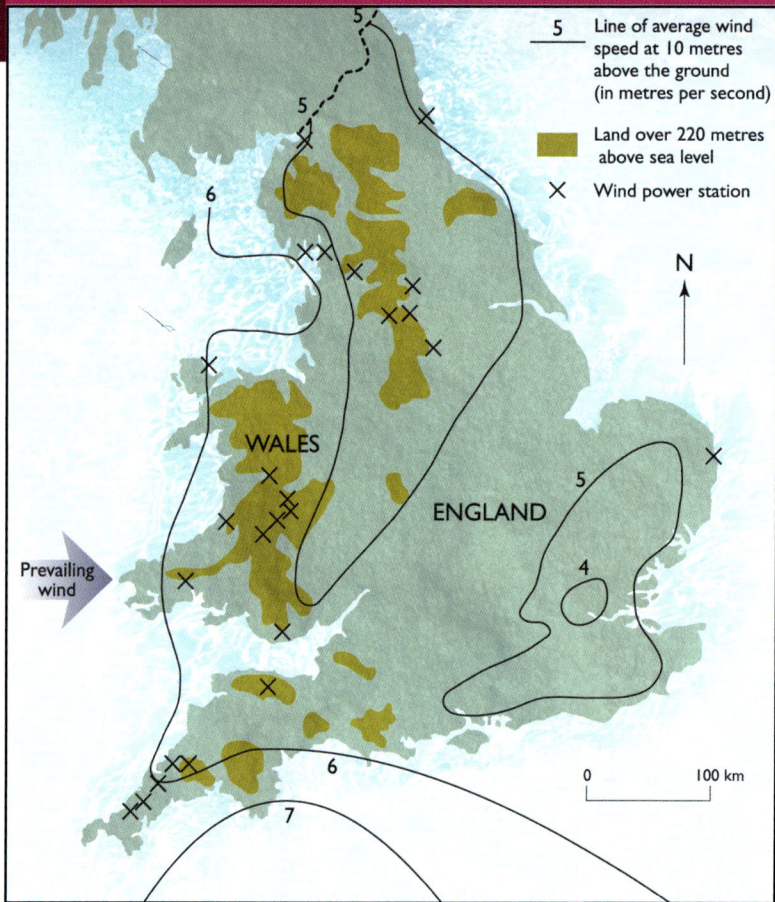

△ **Figure 6.26** Location map for a wind farm

▽ **Figure 6.27** Map of LEDC village

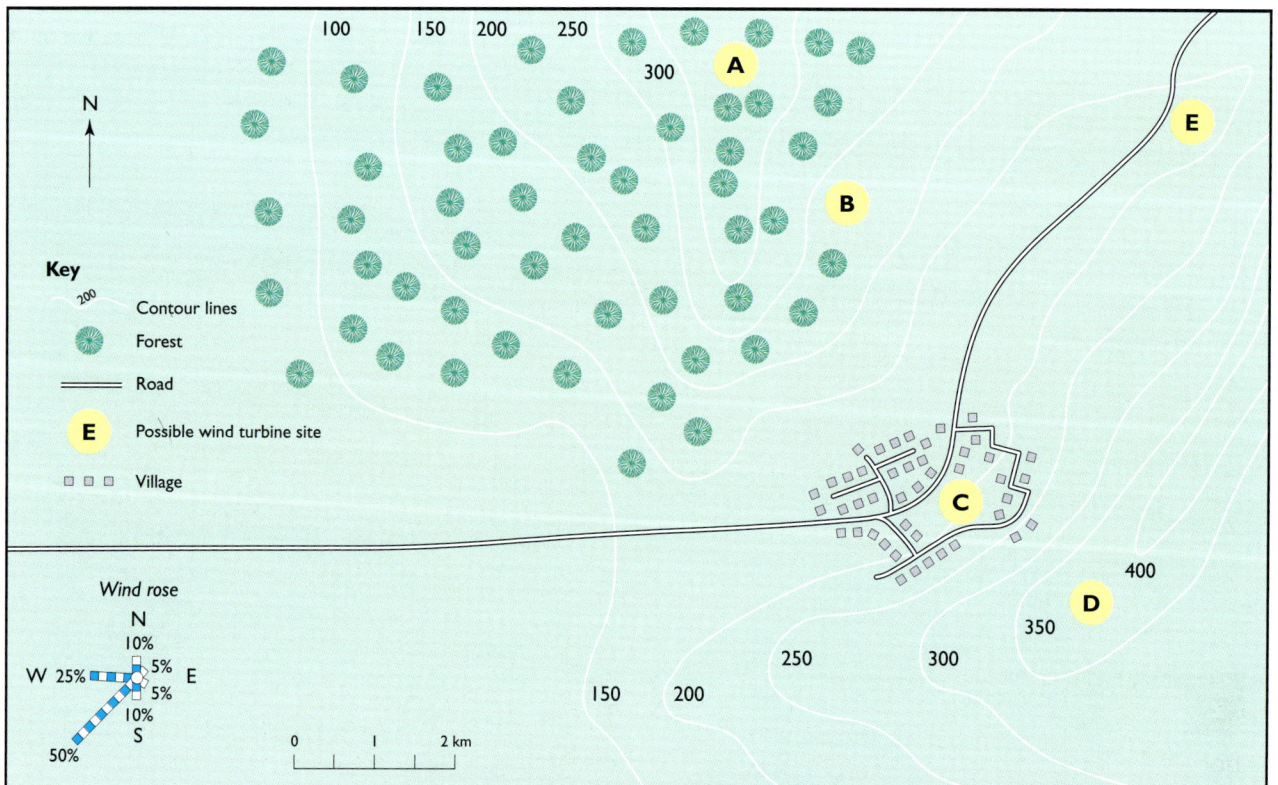

Review

Why is having enough resources important?

Activities

1 The difference between surviving and having a good quality of life is access to basic resources.

a Think back through the work you have done for this chapter. Decide on nine resources that would enable you to survive and have a good quality of life.

b Write each on a piece of card.

c Now arrange these cards as a diamond nine by placing the most important resource at the top and least at the bottom.

most important

least important

2 How would your life be affected if your most important resource were to become exhausted or unavailable?

3 Would your suggested resources be the same as those for someone living in a country at a different level of economic development? Explain your answer.

Glossary

Accessibility
How easy a place is to get to.

Agenda 21
A 1991 international agreement on environment and development.

Aggregate
Broken stone.

Agribusiness
Farming which uses many inputs and is very highly organised, using advanced technology.

Ammonites
Animals like squids in coiled shells living in seas from 280 to 65 million years ago.

Amphibian
An animal that lives on both land and water.

Arch
Curved feature created by erosion process wearing through the rock.

Artificial
Something which is not natural.

Assault
A violent attack on a person.

Attrition
Particles in the water wearing away and rounding each other.

Backwash
Washing sediment back down the beach.

Basalt
A fine grained dark igneous rock.

Bauxite
Mineral containing alumina (source of aluminium).

Bedding planes
Horizontal lines separating one layer of sedimentary rock from another.

Biological weathering
The breakdown by animals and plants, such as root growth and burrowing.

Biomass
The amount of living material in an ecosystem.

Buoys
A floating object that is anchored in place.

Burglary
Entering a building with the intent to steal.

Calcite
The main mineral making up limestone.

Canopy
An almost continuous cover of trees.

Carbonic acid
A weak acid caused by carbon dioxide dissolving in water.

Carboniferous period
From 280 to 345 million years ago.

Caverns
Small underground hollows or caves.

Caves
Large, natural underground hollows.

CCTV
Closed Circuit Television, used for security.

Chalk
A sedimentary rock – a pure white limestone.

Chemical weathering
When water or air alters the chemicals in a rock.

Choropleth map
A map with different shadings showing distributions.

Classify
Organise into groups according to type.

Claustrophobic
Fear of small spaces.

Clay
A sedimentary rock – very fine grains.

Climate
A general look at the weather in a place – what it is typically like.

Coccoliths
Skeletons of sea algae.

Colony
Settlement or settlers in a new country.

Community
A group of people, plants or animals having something(s) in common, e.g. the place where they live.

Commuter
Someone who travels a distance to work, e.g. from a village to a town.

Computer virus
A program infecting computers.

Conflict
A disagreement over how a resource should be used.

Consumption
The process of using up something.

Convenience goods
Goods bought and used frequently such as milk, newspapers.

Coral
A small sea creature which has a hard skeleton.

Corrasion (abrasion)
The stones and sand in the water grinding the rock away.

Country
The area of land that a state occupies.

Cretaceous period
From 65 to 135 million years ago.

Crime wave
A big increase in crime.

Crinoid
A small sea creature with a stem, five arms and a hard skeleton.

Culture
A belief system, or type of social development.

Defensible space
Space that can be more easily protected from crime, e.g. a park overlooked by houses.

Deforestation
Removal of trees.

Desertification
The process by which land becomes desert.

Developing country
A country with a lower economic standard of living.

Development
A complex idea which considers how rich or poor people are. It can be measured in many different ways, including natural, social, economic and political.

Distribution
The spread of something throughout an area.

Documentaries
Factual television programmes.

Domestic
Of the home.

Drought
Continuous dry weather, when water is in short supply.

Dry valleys
Valleys which no longer have a stream running through on the surface.

Dual heritage
Parents from different ethnic groups.

Ecology
Study of the relationships between animals and plants.

Economic output
The amount produced by industry.

Ecosystem
The plants, animals and environment of a particular area, and the links between them.

Energy
The power available to erode, transport and deposit material.

Eroded
Worn away.

Erosion
The wearing away of the landscape by air.

Estuary
Tidal mouth of a river.

Evaporite limestone
Limestone formed when sea water evaporates.

Evolved
Developed by natural processes.

Exports
Goods sold abroad.

Extracted
Removed from the ground using force.

Factor
An explanation for something.

Fallow
Farmland which is given a break from growing crops.

Fear of crime
The fear felt by people about whether they will be a victim of crime.

Fetch
The distance of water across which the wind blows before it meets the coast.

Fjord
A long narrow sea inlet which has steep sides and deep salt water.

Flash floods
Floods which occur quickly without warning.

Flotilla
A fleet of small vessels.

Fluorspar
A mineral.

Fossil
A trace of a life-form found in a rock.

Fossil fuels
Sources of energy formed from the remains of animal and vegetable matter, e.g. oil and coal.

Gated community
A housing area in the USA that ensures visitors pass through security gates.

Genetically modified
Crops which have had their genes changed, e.g. to make them disease resistant.

Geomorphologist
A person who studies the forms and processes of the surface of the earth.

Geomorphology
The study of the shape and the processes that make the landforms of the earth's surface.

Gills
Breathing organ of fish and other water animals.

Glaciated
The word used when landscapes are made by glaciers.

Glacier
A slow moving 'river' of ice.

Global
Worldwide or whole world distributions, e.g. earthquakes, population densities.

Global warming
The rise in the temperature of the earth's atmosphere.

Gorge
Steep-sided narrow valley.

Government
The body of people who organise the running of a state.

Graffiti
Drawings and messages often scribbled or spraypainted in public areas, e.g. a bus shelter.

Granite
An igneous rock.

Graphics
The word used for all drawings and pictures in a computer.

Gross Domestic Product (GDP)
Sum total of a country's output over the year.

Groynes
Wooden barriers pointing out to sea to reduce sand moving along the coast.

Gryke
The gaps between limestone blocks in a limestone pavement.

Habitat
The environment in which an animal or plant lives.

Honeypot
Popular tourist location.

Hospitable
Welcoming.

Humidity
Water vapour content in the atmosphere.

Hydraulic action
Erosion carried out by the sheer power of running water.

Hydro-electric power (HEP)
Electricity produced from running water.

Igneous
Formed from volcanic activity.

Images
Pictures.

Incarceration
Putting people in prison.

Independence
Not depending on another authority.

Indicators
Fact or figure used as evidence.

Inflation
Increase of prices.

Infrastructure
Transport links and basic utilities, e.g. sewage, telephone.

Initiatives
New ideas.

Inner city
The part of a city next to the town centre or Central Business District (CBD).

International
This scale refers to links between two or more countries, e.g. trade.

Irrigation
To supply land with water through artificial channels.

Joints
Vertical cracks in a rock.

Jurassic
From 135 to 195 millions years ago.

Karst
German word for a carboniferous limestone area.

Land degradation
Reducing the quality of the land.

Limestone
Sedimentary rock formed under tropical water.

Limestone pavement
A surface feature made up of blocks of limestone.

Limestone pillar
When a stalactite and stalagmite join up.

Local scale
A unique area about the size of a school catchment area.

Longshore drift
The process of movement of sediment along a coast. This is when waves run at an angle to the coast on a regular basis.

Marble
Metamorphic rock made from limestone.

Marginal
On the edges.

Megawatts
A thousand watts (unit of electricity).

Metamorphic
A rock which has been changed under extreme pressure and/or heat.

Migrant
A person who moves from one place to another to live or work.

Military rule
Ruled by an army.

Minerals
A chemical mixture of elements; rocks are made up of minerals.

Mugging
Robbing someone in a violent way.

Mulatto
Descendant of white and black people.

Nation
People having a common language, history etc., overseen by one government.

National Front
An extremist British political party.

National Park
Protected area of landscape.

National-scale
A whole political unit, e.g. the United Kingdom or a country, e.g. France.

Nation-state
When a nation and a state have the same boundaries.

Natural disasters
Events which occur naturally and cause death/damage.

Neap tide
A tide which has the least difference between high and low water, compared to a spring tide which has the greatest difference between high and low water.

Neighbourhood Watch
Groups of residents meet with the police to discuss ways of reducing local crime.

Newly Industrialising Country (NIC)
A developing country which has shown rapid growth in manufacturing.

Nomadic
Moving from place to place.

Non-recorded crime
Crime not reported to the police and so will not be included in official statistics.

Non-renewable resource
One that will run out because there is a limited amount of it, e.g. oil.

Obese
Someone who is very overweight.

Obsession
Very keen interest.

Ocean currents
The movement of the seas and oceans. Warm water flows to cold areas or vice versa.

Oolitic
A type of limestone with spherical particles.

Organised crime
Crime committed by large groups of people who often treat their activities as a business.

Overseas investments
Investing money in industry or services abroad.

Permeability
How easily water passes through a rock through the cracks and the gaps between the grains.

Permian period
From 225 to 280 million years ago.

Personal scale
An individual's immediate surroundings, i.e. the space close to their body.

Pervious
Water travels through the rock via the bedding planes and joints.

Photosynthesis
Process of using sunlight for plants to generate food.

Physical weathering
The breakdown by wetting and drying, freezing and thawing or heating and cooling.

Pie chart
A way of showing percentages in a diagram.

Pillar
Created when a stalactite and stalagmite join together.

Plantation
Large-scale agricultural production of just one crop, e.g. coffee.

Policy
A course of action to be taken.

Politics
Government and public life.

Population density
Number of people per square kilometre (how crowded).

Porosity
How much water a rock can hold between the grains, like a sponge.

Portland limestone
Limestone found in Dorset used for the Houses of Parliament and other famous London buildings.

Potholers
Activity exploring underground cave systems.

Prevailing wind
The direction in which the wind blows the most often.

Process
The way things happen; in physical geography it refers to erosion, transportation and deposition.

Proteus
Largest cave animal in the world, a 'human fish'.

Quarries
A place where rock is extracted from the ground.

Radiation
Heat waves from the sun.

Raw materials
Natural resources used in manufacturing.

Reality
Something that actually exists.

Recorded crime
Crime reported to the police. This will be included in official statistics.

Reefs
Land made from coral.

Reforestation
Replanting an area with trees.

Region
An area that may be made up of part of one or several countries, that has either physical or human features unique to it.

Relief
The shape of the land's surface.

Renewable energy source
One that will not run out, e.g. the wind.

Resources
A supply of materials that can be used.

Responsibility
Being in charge of good actions.

Restoring
Returning land to its original state.

Right
Just and fair treatment.

Rural
Describing the countryside.

Sandstone
A sedimentary rock made from grains of sand.

Sanitation
The treatment of waste water, e.g. drainage and sewage to protect public health.

Scarce
Hard to find.

Scribe
A writer.

Scenery
A view of the area around you.

Secondary industries
Manufacturing industries.

Sediment
The term used for all sizes of rock material from fine clay to boulders.

Sedimentary
A type of rock made from deposits.

Shoplifter
A person who steals goods from shops.

Site
The piece of land something is located on.

Slavery
Practice of people being owned by other people.

Slope
The angle of the land.

Sloth
Slow-moving mammal with curved claws for gripping trees.

Solution
Water dissolving chemicals in the rocks.

Specialist
A person who is very knowledgeable about a topic.

Spit
A long and narrow bank of sand, shingle and pebbles caused by **longshore drift**.

Spreadsheet
Software which allows you to create tables of data (information and numbers) and do calculations with them.

Springs
Where water emerges from the ground.

Stalactite
A calcium carbonate 'icicle', hanging from the roof of a limestone cave.

Stalagmite
A calcium carbonate column on the floor of a limestone cave.

Standard of living
The quality of our day to day life.

State
An area of land whose people have an independent government.

Subsistence farmer
A farmer who grows food mainly for the family to eat.

Surveillance
Keeping watch over an area.

Sustainable development
Development that does not damage the environment for the future.

Swallow hole
Where water disappears underground.

Swash
Washing sediment up the beach.

Target hardening
Making property more difficult for crime to take place, e.g. fitting security locks.

Temperate
Temperatures that are neither too hot nor too cold.

Territory
Space that you feel belongs to you, e.g. a back garden.

Terrorism
Organised violence, especially for political end.

Tertiary industry
Industry which provides a service, e.g banks, teachers, Police.

Tides
The rise and fall of the sea caused by the gravitational pull of the sun and moon.

Tourists
Visitor to an area.

Trade deficit
Cost of imports is greater than the money made from exports.

Trade surplus
Money from exports greater than the money spent on imports.

Transnational corporation
A large company which operates in more than one country.

Tribes
Group of families with a recognised chief.

Tropical
Located between the Tropic of Cancer and the Tropic of Capricorn.

Undoubtedly
True, without doubt.

Unsanitary
Unhealthy conditions that can cause illness.

Urban
Describing towns and cities.

U-shaped valley
A long narrow valley made by a glacier in a mountainous region. It has a flat bottom with steep sides.

Vandalism
Deliberate damaging of property.

Vertebrate
An animal with a spine.

Waves
The regular up and down movement of the sea caused by the wind blowing over the surface.

Weathered
Attacked by the weather.

Weathering
The breakdown of rocks by the weather, chemical reactions and by plants and animals.